ROUTLEDGE LIBRARY EDITIONS: SCIENCE AND TECHNOLOGY IN THE NINETEENTH CENTURY

Volume 2

ENGLISH MEN OF SCIENCE

ENGLISH MEN OF SCIENCE
Their Nature and Nurture

FRANCIS GALTON

LONDON AND NEW YORK

First published in 1874 by Macmillan & Co.
This edition published in 1970 by Frank Cass & Co. Ltd.
This edition first published in 2019
by Routledge
2 Park Square, Milton Park, Abingdon, Oxon OX14 4RN
and by Routledge
52 Vanderbilt Avenue, New York, NY 10017

Routledge is an imprint of the Taylor & Francis Group, an informa business

© 1874 Francis Galton © New introduction 1970 Ruth Schwartz Cowan

All rights reserved. No part of this book may be reprinted or reproduced or utilised in any form or by any electronic, mechanical, or other means, now known or hereafter invented, including photocopying and recording, or in any information storage or retrieval system, without permission in writing from the publishers.

Trademark notice: Product or corporate names may be trademarks or registered trademarks, and are used only for identification and explanation without intent to infringe.

British Library Cataloguing in Publication Data
A catalogue record for this book is available from the British Library

ISBN: 978-1-138-39006-5 (Set)
ISBN: 978-0-429-02175-6 (Set) (ebk)
ISBN: 978-0-367-07460-9 (Volume 2) (hbk)
ISBN: 978-0-367-07459-3 (Volume 2) (pbk)
ISBN: 978-0-429-02091-9 (Volume 2) (ebk)

Publisher's Note
The publisher has gone to great lengths to ensure the quality of this reprint but points out that some imperfections in the original copies may be apparent.

Disclaimer
The publisher has made every effort to trace copyright holders and would welcome correspondence from those they have been unable to trace.

ENGLISH MEN OF SCIENCE

THEIR NATURE AND NURTURE

BY

FRANCIS GALTON

SECOND EDITION
With a New Introduction by
RUTH SCHWARTZ COWAN

FRANK CASS & CO. LTD.
1970

Published by
FRANK CASS AND COMPANY LIMITED
67 Great Russell Street, London WC1B 3BT

New introduction Copyright © 1970 Ruth Schwartz Cowan

First edition	1874
Second edition	1970

ISBN 0 7146 1622 2

*Printed in Great Britain by Clarke, Doble & Brendon Ltd.
Plymouth and London*

INTRODUCTION TO
THE SECOND EDITION

The twentieth century reader must be cautioned not to laugh at *English Men of Science*. Occasionally the book may seem wondrously, even humorously, lacking in the sophistication that we expect of a scientific classic. But *English Men of Science* is a classic of a different sort: a beginning not an end, the first rather than the last stage in the growth of a science. Francis Galton has been honoured as the founder of biostatistics and as one of the creators of modern psychology. *English Men of Science* is one of his most important works. Parts of it are profound, parts of it are absurd, but taken together they constitute a book that was to be immensely influential.

With the gift of hindsight we can now separate the profound parts from those that are absurd, but for Galton the two were inextricably combined. Today we know that his convictions about the use of statistical techniques in biology and psychology were to prove extremely fruitful, but that his belief in heredity as the predominant force in the production of outstanding intellect was, at best, a naive oversimplification.

Through his efforts, and *English Men of Science* was part of that effort, Galton hoped to establish a body of statistical knowledge about mental heredity. This information would then be united, or so Galton wished, by the eugenic programme of social reform and the end result would be a new pattern of behaviour for society. Each part of this fabric would be useless without the other. It is important to remember, while reading *English Men of Science*, that while the immediate goal of the author was the establishment of certain facts, his ultimate goal was the reformation of society.

Perhaps by tracing the events and ideas that led Galton to the publication of *English Men of Science* (1874) we can understand how his social programme and his scientific interests became so intertwined. Francis Galton was born in 1822, the youngest of seven children of Samuel Tertius Galton, a successful Birmingham banker, and Violetta Darwin Galton, daughter of

Erasmus Darwin. Francis Galton was, therefore, first cousin to Charles Darwin and, like the latter, heir to the Birmingham Quaker scientific tradition. Galton's early years were not unlike those of many another eminent Victorian. First considered something of a prodigy (he learned the alphabet at 18 months, was reading at 2½, signed his name at 3, was memorising poetry at 5 and discussing the Iliad at 6) his school years were, nonetheless, relatively undistinguished. He showed little aptitude for classical studies and chafed at the limitations that were placed upon him, occassionally becoming something of a disciplinary problem. Finally, at the age of 16 he was removed from school, much to his joy, and placed as an indoor pupil at Birmingham General Hospital in preparation for a career as a physician. The next year (1839) found him enrolled at King's College Medical School, but by the end of that year a long-standing wish to read mathematics triumphed over the lures of a medical career and he was allowed (notwithstanding some parental objection) to proceed to Cambridge in 1840.

His career at Cambridge (Trinity College) was also undistinguished. He was an active collegian (Union debates, breakfast parties, drinking societies and the like) but when the time came to prepare for the honours examinations in mathematics he suffered something of a nervous breakdown and was forced to leave the university with a poll degree. Apparently for want of something better to do he spent a few more terms studying medicine, first at Cambridge and then at St. George's Hospital, London, but his desultory progress toward a career was abruptly ended when, in the autumn of 1844, his father died and left him an enormous inheritance.

Medical studies were quickly abandoned and within a few months, "being much upset and craving for a healthier life,"[1] he set out on a tour of the Middle East. The next eight years were his *wanderjahre*: from 1845-1846 he was in the Middle East, from '46-'50 he pursued the sporting life in Scotland, from 1850-1852 under the aegis of the Royal Geographic Society, he led an expedition to South West Africa. Most of his correspondence during these years was destroyed by his family but there is some extant evidence that he suffered a good deal of mental anguish during those years.

At any rate, after returning from Africa in 1852 Galton began to settle down, at least overtly. Within a year he married Louisa

[1] Francis Galton, *Memories of My Life* (London, 1907), p. 82.

Butler, daughter of the headmaster of Harrow, and several months later they took up permanent residence in London. Although physically settled Galton still seems to have been mentally restless—a rich man searching for something meaningful and useful to occupy his time, some cause to which he could commit his overabundant wealth and energy. In 1853 he wrote up an account of his exploration for the Royal Geographic Society (*Tropical South Africa,* 1853) and soon became active on the council of that society. In 1854, at the time of the first British military disasters in the Crimea, he decided to offer his services to the government and spent the next year giving courses in campaigning to the troops at Aldershot, at his own expense. *The Art of Travel,* a guide for travellers who must rough it, was published in the same year. In 1858 he was appointed to the Kew Committee and became much involved in the activities of the observatory. As a result of his work at Kew he became interested in meteorology and in the early '60's published several papers on weather prediction, weather mapping and cyclone behaviour. In 1860 he was made a fellow of the Royal Society, joined the Statistical Society, and a few years later, became involved in the politics of the Ethnological Society.

If all this activity seems to be little more than the random efforts of a man trying to find some useful way to occupy his time, there is at least one common thread that runs through all of his work of this period—an interest in mathematics, especially measurement and counting. There is a certain personality type which is characterized by an unusual if not compulsive, interest in measuring and counting. Galton was a magnificent specimen of the type. He loved to count things, to measure accurately, to measure things that had never been measured before, to invent new and improved devices for measurement. Most of his geographic papers are concerned with accuracy in mapmaking and location finding. In *The Art of Travel* he devoted a chapter to the importance of accurate statistics compiled by travellers. At Kew he was especially involved in testing the accuracy of instruments. His meteorological papers are concerned with more consistent measurement of climatic conditions and mathematical means of prediction from weather- data. During this period he invented a pocket altizimuth and a zeometer, the latter intended to measure the height of mountains. In a more lighthearted vein, he performed experiments to determine mathematically the best combination of variables to produce the best cup of tea (temperature of the water and of the pot, amount of tea and amount of water, size of pot, length of brewing time) and also invented a technique for determining a woman's measurements by triangulation. Galton was, to use a term later coined by his disciple Karl Pearson, a passionate statistician.

The year 1865 finally marked the end of Galton's mental perambulations; for the first time he gave evidence of having found an idea, a vocation, a commitment that could give consistent meaning to his life. In the summer of that year he published a two-part paper in *Macmillan's Magazine* entitled "Hereditary Talent and Character". As its title suggests the paper deals with the problem of whether the mental traits that characterize man can be inherited in the same manner as his physical traits. It is difficult to know precisely what stimulated Galton's interest in this subject. Part of the stimulus may have come from the *Origin of Species*, of which Galton later wrote, "(The) book drove away the constraints of my old superstition, as if it had been a nightmare, and was the first to give me freedom of thought."[2] It is also possible that he was influenced by the work of Quetelet, the Belgian astronomer and meteorologist, who had also attempted to catalogue statistical data about human traits and who had popularised the idea of the statistically average man, *l'homme moyen*.

Whatever the influences that were operating on Galton it is clear that the ideas he enunciated in "Hereditary Talent and Character" set the tone for all his later work. In this paper Galton reported the results of a series of statistical analyses of biographical dictionaries. He had wanted to discover how many of the men listed in those dictionaries had been related to each other and found that the frequency of outstanding men who are related is far greater than the frequency of outstanding men in the general population. From this fact he concluded that mental traits are inherited, and that whatever qualities of mind made men outstanding must have been passed to them by hereditary mechanisms. These qualities of mind could not have been the product of their training or their environment.

Now it may seem to the modern reader that Galton had little definite data on which to build such a sweeping generalisation. The fact that many outstanding men are related to each other

[2] Letter, Francis Galton to Charles Darwin, 26 December 1869. Photographically reproduced in Karl Pearson, *The Life, Letters and Labours of Francis Galton*, Vol. 1 (Cambridge, 1914), p. 6.

does not, of course, prove that their genius was transmitted to them by heredity. For all we know they may have been raised in the same household or taught by the same teacher. Nothing in Galton's statistics gives us reason for believing that one cause (heredity) is stronger than another (environment).

Why did Galton believe in heredity so emphatically? Because in that "fact" he thought he had found a clue to the reformation of society. What is wrong with contemporary society, he said in this paper of 1865, is that the evolution of man has not kept pace with the evolution of technology and social institutions. Marriage customs and enforced celibacy have prevented natural selection from operating on human populations. The ideal society would take advantage of the fact that mental superiority can be bred in much the same way that race horses are bred for speed. An ideal society, he said, would award prizes to marriages likely to produce valuable offspring and would discourage marriages that might produce liabilities.

Galton had finally found a "cause" which could at one and the same time satisfy his scientific curiosity and his social conscience. He did not, however, see the fallacies in his scientific argument because he had become committed to the social programme that was built upon that argument. From 1865 until the end of his life in 1911 almost all of his activities were devoted to implementing his ideas about mental heredity and social reform. What began as a tentative programme in 1865 was amplified in 1869 *(Hereditary Genius)*, reinforced in 1874 *(English Men of Science)*, articulated as "eugenics" in 1883 *(Inquiries into Human Faculty)*, and finally given institutional form as the Eugenics Society in 1901.

In 1872 the French botanist Alphonse de Candolle published a book, *Histoire des sciences et des savants depuis deux siecles*, which directly contradicted Galton's ideas. Candolle, who had read *Hereditary Genius* soon after it was published in 1869, did a statistical study of the lives of outstanding men of science. He found that a very high proportion had come from countries or cities that possessed a moderate climate, a democratic government, a tolerant religious establishment, and an important trade centre. He concluded that Galton was wrong and that environmental factors did indeed play a crucial role in the production of outstanding men.[3]

[3] For extensive correspondence between Candolle and Galton about this matter, see Pearson, *Life of Francis Galton*, Vol. 2, p. 135 ff.

Galton quickly took up Candolle's challenge, in an extraordinary way. In 1874, he circularized British scientists, and asked them to report on the origin of their taste for science.

"It was a daring undertaking, to ask as I did . . . every Fellow of the Royal Society who had filled some important post, to answer a multitude of Questions [sic] needful for my purpose, a few of which touched on religion and other delicate matters. Of course they were sent on the strict understanding that the answers would be used for statistical purposes only. . . . The results of the inquiry showed how largely the aptitude for science was an inborn and not an acquired gift, and therefore apt to be hereditary."[4]

On the pages that follow the reader will be able to judge for himself whether or not Galton had, on this occasion sufficient justification for jumping to the same, rather unwarranted, conclusion he had reached in 1865. However, it may be well to point out another fallacy lurking in these pages, somewhat different from that which mars the paper of 1865. Galton asked the Fellows of the Royal Society whether their taste for science was innate or had developed slowly during their youth. Most respondents answered, "innate". Galton realized at the outset that innateness and inheritance were not necessarily synonymous, but by the end of the book he tended to forget and simply assumed that "innate" meant "inheritance". Here, therefore, one encounters another difficulty: Galton was very free with words. He rarely defined his terms, and when he did do so he violated his definition more often than he adhered to it. "Genius", "talent", "faculty", "innate", "inherited",—these are all complex words, which Galton had a tendency to use with utter abandon.

Those readers who are fascinated by the Victorian frame of mind will find much food for thought in *English Men of Science*. The responses that Galton received, and which he reproduced so assiduously, provide a remarkably good picture of the spectrum of opinions that characterized the Victorian scientific intelligentsia. Particularly interesting in this context are Galton's analysis of the types of religious belief common among scientists, and his respondent's comments on the value of the schooling that they received early in their lives. Even the questionnaire that Galton prepared has something to teach us, for it reveals the extent to which the vocabulary of phrenology and faculty

[4] Galton, *Memories*, pp 291-293.

psychology had found their way into normal English prose, and, one can only assume, the Englishman's view of the mind and himself. Faculty psychology, as propounded by Bain, Broca, Jackson and Spencer, among others, maintained that the mind is composed of separate faculties (or qualities) which are probably localised in specific regions of the brain, and which can definitely be measured either craniologically or through overt behaviour. Many terms that Galton uses—"nervous energy", "size of head", "business acumen", "religious bias"—are categories of faculty psychology. They may seem a little peculiar to us, or perhaps rather vague, but they were perfectly acceptable to the informed Englishman in the second half of the 19th century; indeed, he regarded them as extremely scientific.

After all is said, we must remember that *English Men of Science* represents a very early stage in the development of modern psychology and we must be careful not to criticize Galton overly for the cavalier manner in which he drew up his questionnaire, or for his lack of sophistication in drawing conclusions from it. In 1847 psychological questionnaires were a novelty; before Galton, very few had ever been used. Anecdotal and introspective material was still prevalent in psychological literature, and it would be many years before quantitative techniques would provide more objective data. Indeed, *English Men of Science* represents one of the landmarks in the transition from introspective to objective methods in biological and psychological research. In the future, Galton's statistical, non-anecdotal approach was to prove immensely fruitful for the development of science.

During the next thirty years of his life Galton compiled more and more data on which the new statistical psychology could be based. He established anthropometric laboratories, invented devices for measuring physical abilities, published "family albums" for parents to record the traits of their offspring, and offered a good deal of his own money to prizes for complete sets of data. All this activity was motivated by the same dual commitment that inspired *English Men of Science*: the belief that human traits can be inherited, and that precise statistical analysis of this inheritance will provide society with a valuable tool for reforming itself.

Ruth Schwartz Cowan

New York, June 1969

ENGLISH MEN OF SCIENCE:

THEIR NATURE AND NURTURE.

BY

FRANCIS GALTON, F.R.S.,
AUTHOR OF "HEREDITARY GENIUS," ETC.

London:
MACMILLAN & CO.
1874.

[The Right of Translation and Reproduction is Reserved.]

PREFACE.

I UNDERTOOK the inquiry of which this volume is the result, after reading the recent work of M. de Candolle,[1] in which he analyses the salient events in the history of 200 scientific men who have lived during the two past centuries, deducing therefrom many curious conclusions which well repay the attention of thoughtful readers. It so happened that I myself had been leisurely engaged on a parallel but more extended investigation—namely, as regards men of ability of all descriptions, with the view of supplementing at some future time my work on Hereditary Genius. The object of that book

---

[1] " Histoire des Sciences et des Savants depuis deux Siècles." Par Alphonse de Candolle. Corr. Inst. Acad. Sc. de Paris, &c. Geneve, 1873.

was to assert the claims of one of what may be called the "pre-efficients"[1] of eminent men, the importance of which had been previously overlooked; and I had yet to work out more fully its relative efficacy, as compared with those of education, tradition, fortune, opportunity, and much else. It was therefore with no ordinary interest that I studied M. de Candolle's work, finding in it many new ideas and much confirmation of my own opinions; also not a little criticism (supported, as I conceive, by very imperfect biographical evidence,)[2] of my published views on heredity. I thought it best to test the value of this dissent at once, by limiting my first publication to the same field as that on which M. de Candolle had worked—namely, to the history of men of science, and to investigate their sociology from wholly new, ample, and trustworthy materials. This I have done in the present volume; and I am confident that

[1] Or, "all that has gone to the making of." The word was suggested to me.
[2] Reference may be made to a short review by me of M. de Candolle's work, in the *Fortnightly Review*, March 1873.

one effect of the evidence here collected will be to strengthen the utmost claims I ever made for the recognition of the importance of hereditary influence.

A few of my results, and some of the evidence on which they were based, were given by me at a Friday evening lecture, February, 1874, before the Royal Institution. I have incorporated parts of that lecture into this volume, with emendations and large additions.

It had been my wish to work up the materials I possess with much minuteness; but some months of careful labour made it clear to me that they were not sufficient to bear a more strict or elaborate treatment than I have now given to them.

The pleasant duty remains of acknowledging a debt to my friend, Mr. Herbert Spencer, for many helpful suggestions, and for his encouragement when I was planning this work; and to reiterate my deep sense of gratitude to numerous correspondents, which I have expressed elsewhere in the following pages.

I may add that four of the scientific men who replied to my questions have passed away since I began to write. Of these, two had sent me complete returns, namely, Professor Phillips, the geologist, and Sir William Fairbairn, the engineer. As regards the other two—Sir Henry Holland, the physician, had published his autobiography, but he gave me much help colloquially, and promised more; and Sir Edmund, better known as Count Strzelecki, the Australian traveller and meteorologist, furnished me with very suggestive information, but too incomplete for statistical use.

FRANCIS GALTON.

42 RUTLAND GATE, *November*, 1874.

P.S.—I have to apologise for some faults of style in the earlier pages, due to my not having had as full an opportunity as I had counted upon of correcting that portion of the press.

After I had sent the above to the printer, a friend happened to point out to me the following passage in the "Sartor Resartus" of Carlyle (Bk. ii., ch. 2). It expresses sentiments so nearly akin to those which induced me to write this book, that I am glad to quote it:—

"It is maintained by Helvetius and his set, that an infant of genius is quite the same as any other infant, only that certain surprisingly favourable influences accompany him through life, especially through childhood, and expand him, while others lie close folded and continue dunces. . . . . With which opinion, cries Teufelsdröckh, 'I should as soon agree as with this other—that an acorn might, by favourable or unfavourable influences of soil and climate, be nursed into a cabbage, or the cabbage-seed into an oak. Nevertheless,' continues he, 'I too acknowledge the all-but omnipotence of early culture and nurture: hereby we have either a doddered dwarf bush, or a high-towering, wide-shadowing tree; either a sick yellow cabbage or an edible luxuriant green one. Of a truth, it is the duty of all men, especially of all philosophers, to note down with accuracy the characteristic circumstances of their Education, what furthered, what hindered, what in any way modified it. . . .'"

CONTENTS.

CHAPTER I.

ANTECEDENTS.

Object of book, 1 ; Definition of " Man of Science," 2 ; Data, 10 ; Nature and nurture, 12 ; Race and birth-place, 16 ; Occupation of parents and position in life, 21 ; Physical peculiarities of parents, 27 ; Primogeniture, &c., 33 ; Fertility, 36 ; Heredity, 39. Pedigrees, viz :—Alderson, 41 ; Bentham, 43 ; Carpenter, 43 ; Darwin, 45 ; Dawson Turner, 48 ; Harcourt, 50 ; Hill, 51 ; Latrobe, 54 ; Playfair, 55 ; Roscoe, 57 ; Strachey, 58 ; Taylors of Ongar, 60 ; Wedgewood, 62. Statistical results, 64 ; grandfathers and uncles of scientific men, 65 ; brothers and male cousins, 67 ; Family characteristics, 69 ; Distribution of ability in gifted families, 70 ; Relative influence of paternal and maternal lines, 72.

CHAPTER II.

QUALITIES.

Preliminary, 74 ; Energy, 75, viz :—much above average, 78 ; below average, 97. Size of head, 98 ; Health, 99 ; of parents, 101 ; Perseverance, 103 ; Impulsiveness, 104 ; Practical business habits, 104; Memory, 107; viz.—good verbal, 109; good for facts and figures, 111; for form, 113; good, but no particulars given, 117; bad, 120. Independence of character, 121; ditto of the parents, 122; small religious sects, 123; Mechanical aptitudes, 124; Religious bias, 126; definition of religion, 127; religious sentiment weak, accompanied with scepticism, 130; intellectual interest in religious topics, 130; dogmatic interest, 131; religious bias, 131; ditto with intellectual scepticism, 134. Effect of creed on research, 135, viz.—no deterrent effect, 135; no dread of inquiry, 136; religion and science have different spheres, 136; liberality of early teaching, 137; have early abandoned creeds, 138; creed has had good effect on research, 139; has had some deterrent effect, 140. Truthfulness, 141.

CHAPTER III.

ORIGIN OF TASTE FOR SCIENCE.

Preliminary, 144; Extracts at length, viz:—physics, 149; mathematics, 155; chemistry, 158; geology, 161; zoology, 165; botany, 176; medicine, 180; statistics, 182; mechanics, 184. Analysis of replies, viz.:—tastes strongly innate, 186; decidedly not innate, 191; tastes bearing remotely on science, 194; innate tastes not very hereditary, 196; fortunate accidents, 198; indirect motives or opportunities, 199; professional duties, 202; encouragement at home, 205; influence and encouragement of friends, 211; ditto of tutors, 215; Scotch and English system of tuition, 215; travel in distant parts, 218; unclassed residuum, 221; Summary, 222; Deep movements in national life, 227; Waste of powers, 228; Partial failures, 230; Genius, 233.

CHAPTER IV.

EDUCATION.

Preliminary, 235 ; Merits in education, viz.:—generally praised, 238 ; variety of subjects, 242 ; a little science at school, 243 ; simple things well taught, 243 ; liberty and leisure, 244 ; home teaching and encouragement, 244 ; Merits and demerits balanced, 245 ; Demerits in education, viz.:—narrow education, 246 ; want of system and bad teaching, 251 ; unclassed, 252; Summary, 253; Interpretation of educational needs, 255 ; Conclusion, 258.

APPENDIX.

List of questions sent to scientific men, 261.

ERRATA.

Page 37, line 12, *for* " 30 " *read* " 50."
Page 78, line 4, in the heading, *for* " forty cases " *read* " forty-two cases."

ENGLISH MEN OF SCIENCE.

CHAPTER I.

ANTECEDENTS.

Object of Book—Definition of Man of Science—Data—Nature and Nurture—Race and Birthplace—Occupation of Parents and Position in Life—Physical Peculiarities of Parents—Primogeniture, &c.—Fertility—Heredity—Pedigrees—Statistical Results.

THE intent of this book is to supply what may be termed a Natural History of the English Men of Science of the present day. It will describe their earliest antecedents, including the hereditary influences, the inborn qualities of their mind and body, the causes that first induced them to pursue science, the education they received and their opinions on its merits. The advantages are great of confining the investigation to men of our own period and nation. Our knowledge of them is more complete, and where deficient, it may be supplemented by further inquiry. They are subject to a moderate range of those influences which have the largest disturbing power, and are therefore well fitted for statistical investigation; lastly, the results we may obtain are of direct practical interest. The inquiry is a complicated one at the best; it is advantageous not to complicate it further by dealing with notabilities whose histories are seldom autobiographical, never complete and not always very accurate; and who lived under the varied and imperfectly appreciated conditions of European life, in several countries, at numerous periods during many different centuries.

Definition of "Man of Science."—I do not attempt to define a "scientific man," because no frontier line or *definition* exists, which separates any group of individuals from the rest of their species. Natural groups have nuclei but no outlines; they blend on every side with other systems whose nuclei have alien characters.

A naturalist must construct his picture of nature on the same principle that an engraver in mezzotint proceeds on his plate, beginning with the principal lights as so many different points of departure, and working outwards from each of them until the intervening spaces are covered. Some definition of an ideal scientific man might possibly be given and accepted, but who is to decide in each case whether particular individuals fall within the definition? It seems to me the best way to take the verdict of the scientific world as expressed in definite language. It may be over lenient in some cases, in others it may never have been uttered, but on the whole it appears more satisfactory than any other verdict which exists or is attainable. To have been elected a Fellow of the Royal Society since the reform in the mode of election, introduced by Mr. Justice Grove nearly thirty years ago, is a real assay of scientific merit. Owing to various reasons, many excellent men of science of mature ages, may not be Fellows, but those who bear that title cannot but be considered in some degree as entitled to the epithet of "scientific." I therefore look upon this fellowship as a "pass examination," so to speak, and from among the Fellows of the Royal Society I select those who have yet further qualifications. One of these is the fact of having earned a medal for scientific work; another, of having presided over a learned Society, or a section of the British Association; another, of having been elected on the council of the Royal Society; another, of being professor at some important college or university. These and a few other similar signs of being appreciated by contemporary men of science, are the qualifications for which I have looked in selecting my list of typical scientific men. I have only deviated from these technical rules in two or three cases, where there appeared good reason for their relaxation and where the returns appeared likely to be of peculiar interest. On these principles I drew up a list of 180 men; most of them were qualified on more than one count, and many on several counts. Also, the list appeared nearly exhaustive in respect to those men of mature age who live in or near London, since other private tests suggested few additions. As two of these tests have been proposed by several correspondents, it may be well to describe them. The one is the election of individuals, on account of their scientific eminence, to a certain well-known literary and scientific club, the name of which it is unnecessary to mention. The committee of this club have the power of electing annually, out of their regular turn, nine persons eminent for science, literature, art, or public services. The two or three men who have in each year received this coveted privilege on the ground of science now amount to a considerable number, and they are all on my list. Again, there are certain dining clubs in connection with the Royal Society, the one meeting on the afternoon of every evening that it meets, and the other more rarely, and there are about fifty members to each of these clubs, the same persons being in many instances members of both. The election to either of the clubs is a testimony of some value to the estimation of the scientific status of a man by his contemporaries; almost all their members are on my list. No doubt, many persons of considerable position living in Edinburgh, Dublin, and elsewhere at a distance from London, are not among those with whose experiences I am about to deal. But that is no objection; I do not profess or care to be exhaustive in my data, only desiring to have a sufficiency of material, and to be satisfied that it is good so far as it goes, and a perfectly fair sample. I do not particularly want a list that shall include every man of science in England, but seek for one that is sufficiently extended for my purposes, and that contains none but truly scientific men, in the usual acceptation of that word.

However, I have made some further estimates, and conclude that an exhaustive list of men of the British Isles, of the same mature ages and general scientific status as those of whom I have been speaking, would amount to 300, but not to more.

Some of my readers may feel surprise that so many as 300 persons are to be found in the United Kingdom who deserve the title of scientific men; probably they have been accustomed to concentrate their attention upon a few notabilities, and to ignore their colleagues. It must, however, be recollected that all biographies, even of the greatest men, reveal numerous associates and competitors whose merit and influence were far greater than had been suspected by the outside world. Great discoveries have often been made simultaneously by workers ignorant of each other's labours. This shows that they had derived their inspiration from a common but hidden source, as no mere chance would account for simultaneous discovery. In illustration of this view it will suffice to mention a few of the great discoveries in this generation. That of photography is most intimately associated with the names of Niepce, Daguerre and Talbot, who were successful in 1839 along different lines of research, but Thomas Wedgewood was a photographer in 1802, though he could not fix his pictures. As to the origin of species, Wallace is well known to have had an independent share in its discovery, side by side with the far more comprehensive investigations of Darwin. In spectrum analysis the remarks of Stokes were anterior to and independent of the works of Kirchhoff and Bunsen. Electric telegraphy has numerous parents, German, English and American. The idea of conservation of energy has unnumbered roots. The simultaneous discovery of the planet Neptune on theoretical grounds by Leverrier and Adams is a very curious instance of what we are considering. In patent inventions the fact of simultaneous discovery is notoriously frequent. It would therefore appear that few discoveries are wholly due to a single man, but rather that vague and imperfect ideas, which float in conversation and literature, must grow, gather, and develop, until some more perspicacious and prompt mind than the rest clearly sees them. Thus, Laplace is understood to have seized on Kant's nebular hypothesis and Bentham on Priestley's phrase, "the greatest happiness of the greatest number," and each of them elaborated the idea he had so seized, into a system.

The first discoverers beat their contemporaries

in point of time and by doing so they become leaders of thought. They direct the intellectual energy of the day into the channels they opened; it would have run in other channels but for their labour. It is therefore due to them, not that science progresses, but that her progress is as rapid as it is, and in the direction towards which they themselves have striven. We must neither underrate nor overrate their achievements. I would compare the small band of men who have achieved a conspicuous scientific position, to islands, which are not the detached objects they appear to the vulgar eye, but only the uppermost portions of hills, whose bulk is unseen. To pursue this metaphor; the range of my inquiry dips a few fathoms below the level at which popular reputation begins.

It is of interest to know the ratio which the numbers of the leading scientific men bear to the population of England generally. I obtain it in this way. Although 180 persons only were on my list, I reckon, as already mentioned, that it would have been possible to have included 300 of the same ages, without descending in the scale of scientific position; also it appears that the ages of half of the number on my list lie between 50 and 65, and that about three-quarters of these may be considered, for census comparisons, as English. I combine these numbers, and compare them with that of the male population of England and Wales, between the same limits of age, and find the required ratio to be about one in 10,000. What then are the conditions of nature, and the various circumstances and conditions of life,—which I include under the general name of nurture,—which have selected that one and left the remainder? The object of this book is to answer this question.

DATA.

My data are the autobiographical replies to a very long series of printed questions addressed severally to the 180 men whose names were in the list I have described, and they fill two large portfolios. I cannot sufficiently thank my correspondents for the courteousness with which they replied to my very troublesome queries, the great pains they have taken to be precise and truthful in their statements, and the confidence reposed in my discretion. Those of the answers which are selected for statistical treatment somewhat exceed 100 in number. In addition to these, I have utilized several others which were too incomplete for statistical purposes, or which arrived late, but these also have been of real service to me; sometimes in corroborating, at others in questioning previous provisional conclusions. I wish emphatically to add that the foremost members of the scientific world have contributed in full proportion to their numbers. It must not for a moment be supposed that mediocrity is unduly represented in my data.

Natural history is an impersonal result; I am therefore able to treat my subject anonymously, with the exception of one chapter in which the pedigrees of certain families are given.

NATURE AND NURTURE.

The phrase "nature and nurture" is a convenient jingle of words, for it separates under two distinct heads the innumerable elements of which personality is composed. Nature is all that a man brings with himself into the world; nurture is every influence from without that affects him after his birth. The distinction is clear: the one produces the infant such as it actually is, including its latent faculties of growth of body and mind; the other affords the environment amid which the growth takes place, by which natural tendencies may be strengthened or thwarted, or wholly new ones implanted. Neither of the terms implies any theory; natural gifts may or may not be hereditary; nurture does not especially consist of food, clothing, education or tradition, but it includes all these and similar influences whether known or unknown.

When nature and nurture compete for supremacy on equal terms in the sense to be explained, the former proves the stronger. It is needless to insist that neither is self-sufficient;

the highest natural endowments may be starved by defective nurture, while no carefulness of nurture can overcome the evil tendencies of an intrinsically bad physique, weak brain, or brutal disposition. Differences of nurture stamp unmistakable marks on the disposition of the soldier, clergyman, or scholar, but are wholly insufficient to efface the deeper marks of individual character. The impress of class distinctions is superficial, and may be compared to those which give a general resemblance to a family of daughters at a provincial ball, all dressed alike, and so similar in voice and address as to puzzle a recently-introduced partner in his endeavours to recollect with which of them he is engaged to dance; but an intimate friend forgets their general resemblance in the presence of the far greater dissimilarity which he has learned to appreciate. There are twins of the same sex so alike in body and mind that not even their own mothers can distinguish them. Their features, voice, and expressions are similar; they see things in the same light, and their ideas follow the same laws of association. This close resemblance necessarily gives way under the gradually accumulated influences of difference of nurture, but it often lasts till manhood. I have been told of a case in which two twin brothers, both married, the one a medical man, the other a clergyman, were staying at the same house. One morning, for a joke, they changed their neckties, and each personated the other, sitting by his wife through the whole of the breakfast without discovery. Shakespeare was a close observer of nature ; it is, therefore, worth recollecting that he recognizes in his thirty-six plays three pairs of family likeness so deceptive as to create absurd confusion. Two of these pairs are in the " Comedy of Errors," and the other in "Twelfth Night" (v. 1.) I heard of a case not many years back in which a young Englishman had travelled to St. Petersburg, then much less accessible than now, with no letters of introduction, and who lost his pocket-book, and was penniless. He was walking along the quay in some despair at his prospects, when he was startled by the cheery voice of a stranger who accosted him, saying he required no introduction because his family likeness proclaimed him to be the son of an old friend. The Englishman did not conceal his difficulties, and the stranger actually lent him the sum he needed on the guarantee of his family likeness, confirmed, no doubt, by some conversation. In this and similar instances how small has been the influence of nurture; the child had developed into manhood, along a predestined course laid out in his nature. It would be impossible to find a converse instance in which two persons, unlike at their birth, had been moulded by similarity of nurture into so close a resemblance that their nearest relations failed to distinguish them. Let us quote Shakespeare again as an illustration; in "A Midsummer-Night's Dream" (iii. 2), Helena and Hermia, who had been inseparable in childhood and girlhood, and had identical nurture—

> "So we grew together,
> Like to a double cherry, seeming parted,
> But yet a union in partition,"—

were physically quite unlike: the one was short and dark, the other tall and fair; therefore, the similarity of their nurture did not affect their features. The moral likeness was superficial, because a sore trial of temper, which produced a violent quarrel between them, brought out great dissimilarity of character. In the competition between nature and nurture, when the differences in either case do not exceed those which distinguish individuals of the same race living in the same country under no very exceptional conditions, nature certainly proves the stronger of the two.

RACE AND BIRTHPLACE.

As regards the race of the scientific men on my list, it has already been mentioned that for the purposes of a census enumeration three-fourths may be considered English, but their precise origin is as follows. Omitting a few Germans, out of every 10 scientific men, 5 are pure English ; 1 is Anglo-Welsh ; 1 is Anglo-Irish ; 1 is pure Scotch ; 1 includes Anglo-Scotch, Scotch-Irish, pure Irish, Welsh, Manx and Channel Islands ; finally, 1 is "unclassed." These unclassed are of extremely mixed origin. One is in about equal degrees English, Irish, French, and German; another is English, Scotch-Creole, and Dutch; another English, Dutch, Creole, and Swedish; and so on. (I trust the reader knows what "creoles" are—namely, the descendants of white families long settled in a tropical colony; and that he does not confound the term with "mulattoes.") I give this information without being able to make much present use of it. It is chiefly intended to serve as a standard with which other natural groups may hereafter be compared, such as groups of artists or of literary men.

One would desire to know whether persons in England generally show so great a diversity of origin; but it is somewhat difficult to answer the question owing to a want of precision in the word "generally." If we were to go to rural districts, or small stagnant towns, we should find much less variety of origin; but I think there would be quite as much in the more energetic classes of the metropolis, who have immigrated from all quarters. Some haphazard selecting which I tried confirmed this view. Then comes the important question, Is this a sign that a mixture of one or more of the various civilized races is conducive to form an able offspring? No doubt the varied "nurture" due to separate streams of tradition has great influence in awakening original thought, but we are not speaking of this now; the question is about "nature." On an analysis of the scientific status of the men on my list, it appeared to me that their ability is higher in proportion to their numbers among those of pure race. The Border men and lowland Scotch come out exceedingly well; the Anglo-Irish and Anglo-Welsh, notwithstanding eminent individual exceptions, would as a whole rank last. Owing to my list not being exhaustive, I hardly like to attempt conclusions as to the precise productiveness of scientific ability of the Scotch, English, and Irish severally, but there cannot be a shadow of doubt that its degrees are in the order I have named.

The birthplaces of scientific men and of their parents are usually in towns, away from the sea coast. Out of every 5 birthplaces I find that 1 lies in London or its suburbs; 1 in an important town, such as Edinburgh, Glasgow, Dublin, Birmingham, Liverpool, or Manchester; 1 is in a small town; and 2 either in a village or actually in the country. These returns are given with more detail in the foot-note.[1] The branch of science pursued is often in curious disaccord with the surrounding influence of the birthplace. Mechanicians are usually hardy lads born in the country, biologists are frequently pure townsfolk. Partly in consequence of the prevalence of their urban distribution I find that an irregular plot may be marked on the map of England which includes much less than one-half of its area, but more than 92 per cent. of the birthplaces of the English scientific men or of their parents. The accompanying diagram shows its position; one thin

[1] London, 16; suburbs, 5; = 21. Edinburgh and Glasgow, 7; Cork, Belfast, and Dublin, 6; Birmingham, Liverpool, and Manchester, 5; total = 18. Smaller towns, 21; elsewhere, 40. General total, 100.

arm abuts on the sea between Hastings and Folkestone, and runs northwards over London and Birmingham, where it is joined by another

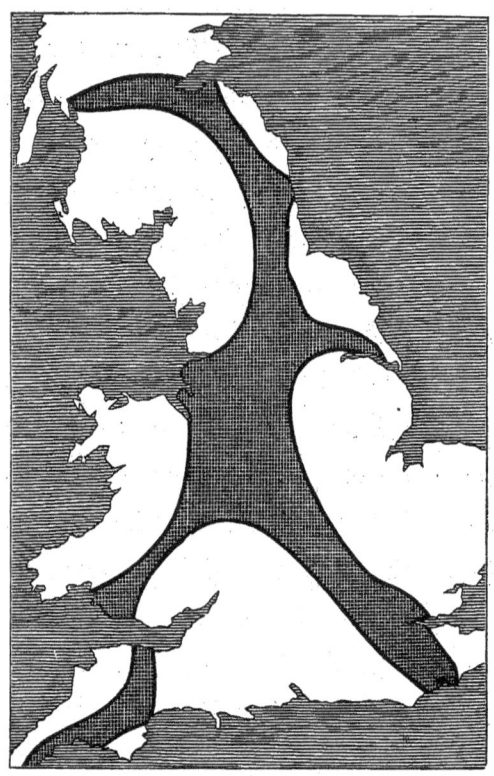

thin arm proceeding from Cornwall and Devonshire, crossing the Bristol Channel to Swansea, and thence to Worcester. The two arms are now combined into one of double breadth; it covers Nottingham, Shrewsbury, Liverpool, and Manchester. Above these latitudes it again narrows, and after sending a small branch to Hull, proceeds northwards to Newcastle, Edinburgh, and Glasgow. Thus there are large areas in England and Wales outside this irregular plot which are very deficient in aboriginal science. One comprises the whole of the Eastern Counties, another includes the huge triangle at whose angles Hastings, Worcester, and Exeter, or rather Exmouth, are situated.

OCCUPATION OF PARENTS AND POSITION IN LIFE.

My list contains men who have been born in every social grade, from the highest order in the peerage down to the factory hand and simple peasant, but the returns which I shall discuss do not range quite so widely. These are 96 in number, and may be classified as follows—but the same name appears in two classes on eleven occcasions, so that the total entries are raised to 107:—

Noblemen and private gentlemen	9
Army and navy, 6; civil service, 9; subordinate officers, 3	18
Law, 11; medical, 9; clergy and ministers, 6; teachers, 6; architect, 1; secretary to an insurance office, 1	34
Bankers, 7; merchants, 21; manufacturers, 15	43
Farmers	2
Others	1
	107

The terms used in the third and fourth groups must be understood in a very general sense; thus, there are some "merchants" on a very small scale indeed, and others on a very large one.

It is by no means the case that those who have raised themselves by their abilities are found to be abler than their contemporaries who began their careers with advantages of fortune and social position. They are not more distinguished as original investigators, neither are they more discerning in those numerous questions, not strictly scientific, which happen to be brought before the councils of scientific societies. There can be no doubt but that the upper classes of a nation like our own, which are largely and continually recruited by selections from below, are by far the most productive of natural ability. The lower classes are, in truth, the "residuum."

Of the 6 clergymen or ministers who were fathers of scientific men, no less than 4 appear in a second category, viz., (1) clergyman and schoolmaster; (2) physician, afterwards clergyman; (3) Unitarian minister and schoolmaster; (4) professor of classics, afterwards an Independent minister. Among the successful graduates of Oxford and Cambridge, and among purely literary men, we find a much larger proportion of sons of clergymen. There is at Cambridge a well-known university scholarship, called the "Bell," which is open only to sons of clergymen of the Church of England. As it has been chiefly given for classical proficiency, we may be almost sure that the senior classic of his year, if he were the son of a clergyman, would also be a Bell scholar. I looked through the lists, and found that out of 45 senior classics (1824–68 inclusive) 10 had gained the scholarship, whence I conclude that at least 1 out of every 4 or 5 Cambridge graduates is the son of a clergyman. At this rate, out of 100 Cambridge graduates, 22 would have had clergymen of the Church of England for their fathers, whereas out of 100 scientific men only 3 or 4 were so circumstanced. It is therefore a fact, that in proportion to the pains bestowed on their education generally, the sons of clergymen rarely take a lead in science. The pursuit of science is uncongenial to the priestly character. It has fallen to my lot to serve for many years on the councils of many scientific societies, and, excepting a very few astronomers and mathematicians, about whom I will speak directly, I can only recall 3 colleagues who were clergymen; curiously enough, 2 of these, the Revs. Baden Powell and Dunbar Heath, have been prosecuted for unorthodoxy; the third was Bishop Wilberforce, who can hardly be said to have loved science; he rarely attended the meetings, but delighted in administration, and sought openings for indirect influence. The reason for the abstinence of clergymen from scientific work cannot be that they are too busy, too much home-tied, or cramped in pecuniary means, because other professional men, more busy, more at the call of others, and having less assured revenues, are abundantly represented on all the council lists.

Not caring to trust my unaided recollections, I have examined the council lists of ten scientific societies at or near the three periods, 1850, 1860, 1870. There have been changes in some of the societies, and there are many trifling peculiarities of detail, tedious and unnecessary here to deal with, but the following statement is substantially correct. The ordinary members of council are on a rough general average 20 in number to each of the following societies : (1) Royal; (2) British Association; (3) Astronomical; (4) Chemical; (5) Geological; (6) Linnæan; (7) Zoological; (8) Geographical; (9 and 10) the two predecessors of the recently-established Anthropological Institute, viz. Ethnological and Anthropological; (11) Statistical.

Therefore as we are dealing with 3 distinct periods, 11 societies, and 20 members of council to each, there have been about $(3 \times 11 \times 20 =)$ 660 separate appointments. Clergymen have held only 16 of these, or 1 in 40; and they have in nearly every case been attached to those subdivisions of science which have fewest salient points to scratch or jar against dogma. Thus Prof. Challis, Dr. Lloyd, Dr. Robinson, Dr. Whewell, Rev. J. Fisher, Rev. W. Webb, Rev. Vernon Harcourt, Prof. Pritchard, Prof. Price, Rev. J. Barlow, and Prof. Willis are all chiefly connected with astronomy, physics, and mathematics; the five remaining names are those of the Rev. G. C. Renouard, the geographer; Bishop Wilberforce, and the Rev. Dunbar Heath, of whom I have already spoken; the Rev. Dr. Nicholson, and the Rev. Canon Greenwell: there is not a single biologist among them.

PHYSICAL PECULIARITIES OF PARENTS.

It has been frequently asserted that certain physical peculiarities in the parents clash, and that others combine happily in the offspring. I therefore thought it well to make inquiries as to the figure, complexion, colour of hair, height, and other physical peculiarities of the fathers and mothers of the scientific men. I also asked about the temperaments, if they were marked, but the answers to these were few.

Tables showing the number of cases in which there has been harmony, indifference, or contrast, between various physical peculiarities of the two parents

TEMPERAMENTS OF PARENTS.

(h = harmony, c = contrast).

MOTHERS.	FATHERS.			
	Nervous.	Sanguine.	Bilious.	Lymphatic.
Nervous	h. 6	5	—	c. 0
Sanguine	1	h. 3	—	c. 0
Bilious	4	—	h. 1	—
Lymphatic	c. 0	c. 2	—	h. 0

Summary—Harmony, 10 cases; contrast, 2; indifferent, 10. Total, 22.

COLOUR OF HAIR OF PARENTS.

(h = harmony, c = contrast).

MOTHERS.	FATHERS.						
	Black.	Dark.	Dark Brown.	Brown.	Light Brown.	Light.	Fair.
Black . . .	h. 2	h. 2	h. 1	1	c. 0	c. 1	c. 0
Dark	h. 2	h. 5	h. 1	2	2	c. 1	c. 1
Dark Brown .	0	h. 2	h. 4	h. 3	3	0	c. 0
Brown . . .	3	4	h. 2	h. 14	h. 1	0	1
Light Brown .	c. 0	2	2	h. 1	h. 0	h. 0	0
Light. . . .	c. 3	c. 0	0	2	h. 0	h. 2	h. 0
Fair	c. 0	c. 0	c. 0	0	h. 1	h. 0	h. 1

Summary—Harmony, 44 cases; contrast, 6; indifferent, 22.
Total, 72.

I have, in addition, 11 cases of coloured hair—yellowish, sandy, red, light auburn, dark auburn, chestnut—but not one case of strict harmony among them.

FIGURE OF PARENTS OF SCIENTIFIC MEN.

(h = harmony, c = contrast).

MOTHERS.	FATHERS.				
	Corpulent, stout, or plump.	Muscular, robust, strong.	Compact, symmetrcl, stately.	Spare, neat, small.	Medium.
Corpulent, stout, or plump . .	h. 3	h. 5	0	c 7	c. 1
Muscular, robust, strong . . .	h. 0	h. 2	1	c. 1	0
Compact, symmetrical, stately .	3	2	h. 2	6	0
Spare, neat, small	c. 9	c. 5	4	h. 12	1
Medium . . .	0	1	1	5	h. 0

Summary—Harmony, 24 cases; contrast, 23; indifferent, 24.
Total, 71.

The foregoing tables show results bearing on the question whether harmony or contrast prevails in the physical characteristics of the parents. I think they must be accepted as decidedly in favour of harmony. The grand totals which they give are 78 cases of harmony, 31 of contrast, and 56 of indifference. In short, there is more purity of breed in scientific men than would have resulted from haphazard marriages. In the temperaments of their parents, harmony strongly prevails over contrast, the proportion being 5 to 1 in favour of the former. In colour of hair, harmony is twice as frequent as contrast. In figure it is equally common, because "corpulent, stout, or plump" persons of one sex seem to have a peculiar and reciprocated liking for "spare, neat, or small" persons of the other. This is literally the only case in these tables where a love of contrast equals that of harmony. I came to much the same conclusions by giving appropriate marks for harmony, contrast, and indifference to each quality in each case, thus obtaining aggregate marks for every pair, which I treated on much the same principle that the separate qualities are treated in the table. As regards height, there is a stricter method of investigation, which statisticians will appreciate. It is well known, by repeated experience, that the heights of men and of women in any large group are distributed according to the "law of frequency of error." In other words, the proportionate number of people of different heights corresponds to what would have been the case supposing stature to be due to the *aggregate action of many small and independent variable causes.* The probability is inconceivably small that all the independent causes should in any given case co-operate to produce an excess of height; if they did so, the result would be a Brobdignagian giant; or that they should all co-operate to produce a deficiency in height, in which case the result would be a Lilliputian dwarf. On the other hand, the probability is great that the number and effects of the causes in excess and those in deficiency of their several average values will be pretty equal. As for these and all other intermediate cases, their relative frequency is determined by the above law, which is based on that by which the relative frequency of different "runs of luck" is calculated.

I now proceed to apply this law. I have 62 cases in which the heights of both parents are given numerically, whence it appears that—(1) the average height of the fathers is between 5 ft. 9 in. and 5 ft. $9\frac{1}{4}$ in., and that their distribution conforms closely to the law of frequency of error, the "probable error" of the series being 1·7 in. (2) The average height of the mothers is 5 ft. $4\frac{1}{2}$ in., and the distribution of their heights conforms fairly to the above-mentioned law, the "probable error" of the series being 1·9 in. It follows from the well-known properties of the law in question, that if there had been no sexual selection in respect of height, the sum of the heights of the two parents would also conform to the law of frequency of error, and that the probable error of the series would be $\sqrt{(1\cdot7)^2+(1\cdot9)^2} = 2\cdot5$ in. (3) I find that the heights in question do conform pretty closely to the law in question, and that the probable error of the series is 2·3 in., which differs so slightly from the value obtained by calculation, on the supposition of there having been no sexual preference for contrast in height, that we may safely affirm in this case also, that the love of contrast does not prevail over that of harmony.[1]

It is a question of high importance to speculations on the future of our race, whether the instincts of sexual selection are or are not repugnant to an improvement in the human breed. We know perfectly well that they are repugnant to unions where the resemblance is very close; thus near intermarriages shock our feelings, and the maintenance of high-bred artificial varieties in their purity is always effected with difficulty among animals. On the other hand, they are equally repugnant to unions in which there is great contrast; thus, the intermarriage of white and black races rarely takes place, and animals of different species refuse to cross. Where, then, and how wide, is the belt that lies

[1] The series of facts in (1), (2), and (3), and the corresponding figures given by the theory with which they are supposed to conform, are as follows:—

	(1) Father.	(2) Mother.	(3) Both Parents.
Fact	3 15 29 30 18 3 2	5 14 32 29 11 6 3	3 18 34 26 18 5 1
Theory	5 15 27 29 18 5 1	8 18 25 26 15 6 2	6 18 31 29 13 2 1

between close harmony and wide contrast, in which sexual instinct acts most powerfully? It appears from the facts in this chapter, that the marriages of parents of the scientific men on my list actually tended to produce differentiation and purity of race. My data concerning the parents of men of other groups are insufficient to enable me yet to give comparative results showing how far the selective sexual instincts of the population generally would thwart, be indifferent to, or co-operate with the influences of future social restrictions on unsuitable marriages, or encouragement of suitable ones.

PRIMOGENITURE, &C.

The following statement shows, in percentages, the position of the scientific men in respect to age among their brothers and sisters:—

Only sons, 22 cases; eldest sons, 26 cases; youngest sons, 15 cases. Of those who are neither eldest nor youngest, 13 come in the elder half of the family; 12 in the younger half; and 11 are exactly in the middle. Total, 99.

It further appears that, at the time of the

birth of the scientific men, the ages of their fathers average 36 years, and those of their mothers 30. The details are shown in the table below:—

No. of Cases.	Age of Parents at Birth of Scientific Men.								Total Cases.
	Under 20	20–	25–	30–	35–	40–	45–	50 and above.	
Fathers	0	1	15	34	22	17	7	4	100
Mothers	2	20	26	34	12	5	1	—	100

Putting these facts together, viz.—(1) that elder sons appear nearly twice as often as younger sons; (2) that, as regards intermediate children, the elder and younger halves of the family contribute equally; and (3) that only sons are as common as eldest sons, we must conclude that the age of the parents, within the limits with which we chiefly have to deal, has little influence on the nature of the child; secondly, that the elder sons have, on the whole, decided advantages of nurture over the younger sons. They are more likely to become possessed of independent means, and therefore able to follow the pursuits that have most attraction to their tastes; they are treated more as companions by their parents, and have earlier responsibility, both of which would develop independence of character; probably, also, the first-born child of families not well-to-do in the world would generally have more attention in his infancy, more breathing space, and better nourishment, than his younger brothers and sisters in their several turns.

The opposing disadvantage of primogeniture, in producing less healthy children and half as many idiots again as the average of the rest of the family, has not been sensibly felt, partly because the latter risk is very small, and partly because the mothers of the scientific men are somewhat less youthful than those from whom the above statistical results were calculated. (See Duncan " On Fertility," &c., second edition, pp. 293, 4, for tabulations of Dr. A. Mitchell's results.) An unusual number of the mothers of the scientific men were between 30—34 at the time of their birth; this is a very suitable age, according to the views of Aristotle, but undoubtedly older than what Dr. Duncan's statistics (pp. 387, 390) recommend. According to these, the most favourable period for the survival of mother and child, and therefore probably the best in every sense, is when she is 20—25 at the time of giving birth. The important question of the effect of the age of the parent on the wellbeing of the offspring seems never yet to have been treated as strictly and as copiously as it deserves. Dr. Duncan, in the chapter of his work above referred to, has discussed the materials at his disposal with great ingenuity and industry; but adequate statistics, sorted according to the various classes of society, are still wanting.

FERTILITY.

The families are usually large to which scientific men belong. I have two sets of returns—the one of brothers and sisters, excluding, for the most part, those who died in infancy; and the other of brothers and sisters who attained 30 years. In these several cases I have included the scientific man himself, and find, on an average of about 100 cases, that the total number of brothers and sisters is 6·30 in the first case, and 4·80 in the second. It is a matter of great interest to compare with these figures the number of the children of the scientific men themselves. It is easy to do so with fairness, because the time of marriage proves to be nearly the same in both cases; if anything, the scientific men marry earlier than their parents. It remains to eliminate all cases of absolutely sterile marriages on the part of the scientific men, and those in which there might yet be other children born. Having attended to these precautions, I find the number of their *living* children (say, of ages between 5 and 30) to be 4·7. This implies a diminution of fertility as compared with that of their own parents, and confirms the common belief in the tendency to an extinction of the families of men who work hard with the brain. On the other hand, I shall show that the health and energy of the scientific men are remarkably high; it therefore seems strange that there should be a falling off in their offspring. I have tried in many ways to find characteristics common to those scientific men whose families were the smallest, but have only lighted upon one general result, which I give provisionally, namely, that a *relative* deficiency of health and energy, in respect to that of their own parents, is very common among them. Their absolute health and energy may be high, far exceeding those of people generally; but I speak of a noticeable falling off from the yet more robust condition of the previous generation : it is this which appears to be dangerous to the continuance of the race. My figures give the remarkable result that there are no children at all in one out of every three of these cases. I think that ordinary observation corroborates this conclusion, and that those of my readers who happen to have mixed much in what is called intellectual society will be able to recall numerous instances of persons of both sexes, but especially of women, possessed of high gifts of every kind, including health and energy, but of less solid vigour than their parents, and who have no children. I do not overlook the fact that the scientific men are an urban population, being mindful of results I have published elsewhere (*Statistical Journal*, 1873), which show a similar diminution in the average fertility of townsmen as compared with country folk ; but this would not account for their being less prolific than their parents who were also townsmen, nor for the large number of wholly sterile marriages.

HEREDITY.

The effects of education and circumstances are so interwoven with those of natural character in determining a man's position among his contemporaries, that I find it impossible to treat them wholly apart. Still less is it possible completely to separate the evidences relating to that portion of a man's nature which is due to heredity, from all the rest. Heredity and many other co-operating causes must therefore be considered in connection; but I feel sure that as the reader proceeds, and becomes familiar with the variety of the evidence, he will insensibly effect for himself much of the required separation. Also, from time to time, as opportunity may offer, I shall attempt to draw distinctions.

The study of hereditary form and features in combination with character promises to be of much interest, but it proves disappointing on trial, owing to the impossibility of obtaining good historical portraits. The value of these is further diminished by the passion of distinguished individuals to be portrayed in uniforms, wigs, robes, or whatever voluminous drapery seems most appropriate to their high office, forgetting that all this conceals the man. The practice might well become common of photographing the features from different points of view, and at different periods of life, in such a way as would be most advantageous to a careful study of the lineaments of the man and his family. The interest that would attach to collections of these in after-times might be extremely great.

PEDIGREES.

Thirteen families have been selected, out of those to which about 120 of the scientific men on my list belong, as appearing noteworthy for their richness in ability during two, three, or more generations, or for any other peculiarity; in some cases they are also remarkable for purity of type. The facts may for the most part be verified by reference to the publications of which the titles are given; and the whole could have been obtained by any one who cared to search other more or less public sources of information. Five of these families (Bentham, Darwin, Dawson-Turner, Roscoe, and Taylor of Ongar) have already been alluded to in my previous work ("Hereditary Genius"), whence I have extracted what appeared to the point, adding what was necessary. In estimating the number of individuals in each generation, the practice has been usually adopted of not counting those who died young, or have not yet attained their 30th year.

ALDERSON.—Many members of this family have been intellectually gifted. There has been an unusual number of cases of mathematical achievement among them.

First generation.—5 males and 2 females, children of the Rev. J. Alderson and his wife (the latter lived to 94). Of these, 3 males deserve notice :—(1) James Alderson, M.D., of Norwich; (2) Robert Alderson, Recorder of Norwich, Ipswich, and Yarmouth; (3) John Alderson, founder and president of all the literary and scientific institutions of the time in Kingston-upon-Hull. All these were men of considerable local repute.

Second generation.—15 males and 12 females, of whom 5 males and 1 female deserve especial mention:—(1) Sir Edward Hall Alderson, Baron of the Exchequer, who was the first man of his year at Cambridge, both in mathematics and classics, being senior wrangler and senior classical medallist, a distinction barely equalled in the long annals of university achievement; (2) Robert Woodhouse, also a senior wrangler, Lucasian and Plumian Professor of Astronomy at Cambridge; (3) the Rev. Samuel H. Alderson, third wrangler, and tutor of Caius College; (4) Sir James Alderson, M.D., F.R.S. (sixth wrangler), for four years President of the Royal College of Physicians; (5) Colonel Ralph Alderson, R.E., a distinguished officer, and one of the first government commissioners of railways; (*1*) Mrs. Amelia Opie, the novelist.

Third generation.—I have not sufficient information, although I know that it includes many persons of ability, among whom is Major H.

Alderson, R.A., a distinguished officer; also a married lady of high artistic powers.

BENTHAM.—A family consisting of only 3 male representatives, all eminent, and one illustrious.

First generation.—2 brothers:—(1) Jeremy Bentham, jurist of the highest rank (life by Sir J. Bowring, prefixed to the collected works edited by him); (2) General Sir Samuel Bentham, whose early manhood was spent in the Russian service; distinguished for his numerous administrative reforms and singular inventive power. Afterwards inspector-general of naval works in England (life by his widow, 1862).

Second generation.—1 male only:—George Bentham, F.R.S., systematic botanist of the highest rank; in early life, writer on logic; for many years President of the Linnæan Society.

CARPENTER.—Among the characteristics of this family are literary and scientific enterprise, philanthropic effort, nonconformity, and aptitude for oral exposition.

First generation.—Rev. Lant Carpenter, LL.D.,

Unitarian minister; descended from a non-subscribing Presbyterian family, and married to a wife of similar descent; a leading member of the Liberal party in Exeter and Bristol; extremely active in the promotion of philanthropic objects; both literary and scientific in his studies, and a man of local celebrity (memoirs by his son, 1842).

Second generation.—2 males and 3 females, of whom both the males and 1 female require notice :—(1) William B. Carpenter, F.R.S., Registrar of the London University, physiologist, and frequent writer and speaker on scientific subjects, in many cases connected with social amelioration; (2) Dr. P. P. Carpenter (of Montreal), conchologist; actively engaged in philanthropic work; (*1*) Mary Carpenter, actively engaged in the foundation and organization of philanthropic institutions, especially juvenile reformatories, and promoter of female education in India.

Third generation (too young for special notice) includes an influential dissenting minister and a very successful student.

DARWIN.—There are many instances in this family of a love for natural history and theory, and of an aptitude for collecting facts in business-like but peculiar ways. Speaking from private sources of knowledge, I am sure that these characteristics are hereditary rather than traditional; there is also a strong element of individuality in the race which is adverse to traditional influence.

First generation.—(1) Erasmus Darwin, M.D., F.R.S., physician, physiologist and poet. His "Botanic Garden" had an immense reputation at the time it was written; for, besides its intrinsic merits, it chimed in with the sentiments and mode of expression of his day. The ingenuity of Dr. Darwin's numerous writings and theories is truly remarkable. He was held in very high esteem by his scientific friends, including such celebrities as Priestley and James Watt, and it is by a man's position among his contemporaries and competitors that his worth may most justly be appraised. Unfortunately for his memory, he has had no good biographer. He was a man of great vigour, humour and geniality (Miss Seward's life of him, and latterly a pamphlet by Dr. Richardson;

see also Meteyard's "Life of Wedgewood"); (2) his brother, Robert Waring Darwin, wrote "Principia Botanica," which reached its third edition in 1810. It is said (Meteyard's "Life of Wedgewood") that the Darwins "sprang from a lettered and intellectual race, as his (Dr. Darwin's) father was one among the earliest members of the Spalding Club."

Second generation.—7 males, 3 females, of whom 3 males deserve notice:—(1) Charles Darwin, who died at the age of only 21, poisoned by a dissection wound, but who had already achieved such distinction that his name has been frequently mentioned in biographical dictionaries. His thesis, on obtaining the gold medal of the Edinburgh University, was on the distinction between " pus " and " mucus." It was a real step forward in those early days of exact medical science, and was thought highly of at the time; (2) Robert Waring Darwin, M.D., F.R.S., a physician, and shrewd observer, of great provincial celebrity, on many grounds, who lived at Shrewsbury. He married a daughter of Wedgewood's, and was father of Charles Darwin (see below); (3) Sir

Francis Darwin, originally a physician, but for many years living in a then secluded part of Derbyshire, surrounded by animal oddities; half-wild pigs ran about the woods, tamed snakes frequented the house, and the like.

Third generation.—8 males, 14 females, of whom 3 males may be mentioned; but illustriously among them—(1) Charles Darwin, F.R.S., "the Aristotle of our days," whom all scientific men reverence and love; the simple grandeur of whose conclusions is as remarkable as the magnitude and multifariousness of their foundation. There is much ability in many individuals in this generation who bear the name of Darwin, and it has been strongly directed to natural history in the case of (2) a son of Sir Francis Darwin, a frequent writer, under a well-known *nom de plume*, on sporting matters. Among those who do not bear that name (being children of the daughters of Dr. Erasmus Darwin), I mention (3) myself,[1]

---

[1] Captain Douglas Galton, F.R.S., distinguished for official activity in many high posts, and now Director of Public Works, is descended maternally, not from the Darwin, but from the Strutt family, which has produced noted mechanicians.

with all humility, as falling technically within the limits of the group of scientific men under discussion, on the ground of former geographical work, and having had much to do in the administration of various scientific societies.

Fourth generation.— Includes very few individuals who have reached mature manhood; among these are (1) George Darwin, second wrangler at Cambridge, author of an important article on " Restrictions to Liberty of Marriage ;" (2) Captain Leonard Darwin, R.A., who was second in the competition of his year for Woolwich, and now engaged on the Transit of Venus Expedition; (3) Henry Parker, fellow of University College, Oxford, classical scholar and chemist.

DAWSON-TURNER.—This family is characterised by great intellectual activity and much artistic taste.

First generation.—Dawson Turner, F.R.S., botanist, scholar, antiquary; a man of unwearied activity in collecting and compiling, and an encourager of work in others. One of his two uncles was the Rev. Joseph Turner, senior wrangler in 1768, and much distinguished by the personal friendship of Mr. Pitt. Among his 10 male first cousins on the paternal side were the late Lord Justice Turner and his accomplished brothers.

Second generation.—2 males and 6 females. The latter were all remarkable for their energy, accomplishments, and the large share they took in the literary labour of their father and husbands, which was not confined to transcribing. Three were accomplished artists, one a musician, another well versed in Greek.

Third generation.—Of those above the age of 30 there are 5 males and 3 females, of whom 4 males deserve mention :—(1) Dr. Joseph Hooker, president of the Royal Society, very eminent botanist, director of Kew Gardens, and formerly Thibetan traveller, and naturalist to an antarctic expedition ; his father was Sir William Hooker, F.R.S., also one of the first botanists of his day, and director of Kew Gardens; (2) Francis Palgrave, editor of the "Golden Treasury," scholar and art critic ; (3) Gifford Palgrave, orientalist, Arabian explorer, and author of one of the most remarkable works of travel ever written ; (4) R. H.

Inglis Palgrave, statistician. (The father of the three last was Sir Francis Palgrave, historian.)

HARCOURT.—Scholastic success, with much love for science.

First generation.—The Rev. Vernon Harcourt, archbishop of York; a man of polished intellect and social gifts.

Second generation.—10 males and 3 females, of whom 4 males deserve notice :—(1) The Rev. W. Vernon Harcourt, F.R.S., chemist, the first president and one of the founders of the British Association at a time when science was partly ridiculed and partly denounced. He was the chief framer of its elaborate constitution, which is, I believe, a solitary instance of the invention of a complex administrative machinery which worked perfectly from the first, and has continued working, almost unchanged, for nearly half a century. It has served as a model upon which many other societies have organized themselves. (2) Egerton; and (3) Edward Vernon Harcourt, both double-firsts at Oxford; and (4) Granville Vernon Harcourt, who died when an undergraduate at Oxford, having gained the Latin university prize.

Third generation.—10 males and 13 females, of whom 2 males deserve mention:—(1) Sir William Vernon Harcourt, M.P., lately solicitor-general, professor of international law at Cambridge, well known as a political writer under the name "Historicus"; (2) Augustus G. Vernon Harcourt, F.R.S., a distinguished chemist, Lee's reader in chemistry at Oxford.

HILL.—The characteristics of this family are, active interest in social improvement, power of organization, mechanical aptitude, and general sterling worth. Its type in the second generation seems to have been unusually pure.

First generation.—Thomas Wright Hill, descended from stanch Independents, and married to a wife of equal vigour and fortitude, who came from a family noted for mechanical aptitude, which she transmitted to her descendants. He rose by his own exertions, and (æt. 40) established a school, much spoken of at the time, on an entirely new principle of management at Hazelwood, near Birmingham. The boys were taken into administrative co-operation; they regulated their own discipline, and the things they learnt were of the most varied kind. Some men of high note were educated there, and, among these at least one of the scientific men on my list. He gave much attention to mental calculation, and even on his deathbed (æt. 88) invented and successfully applied a new method for determining for any year the date of Easter. Also known for his analysis of articulate sounds and phonography. (Short biographical notice in Annual Report R. Astronomical Society, Feb. 13, 1852.)

Second generation consisted of 5 males and 2 females.—All 5 males had strong points of resemblance and deserve notice. (1) Sir Rowland Hill, K.C.B. and F.R.S., originator and organizer of the system of penny postage, which is an influence of the first order of magnitude in modern civilization. He was noted in youth for powers of mental calculation, and in some points was superior even to Zerah Colburn and George Bidder; thus he could mentally extract to the nearest integer the cube root of any number not exceeding two thousand millions. First inventor (1835) of rotatory printing, the method which, with slight changes of detail, is still in use for newspapers. Rewarded by three separate grants, viz., in 1846 by a public testimonial of the value of 13,360*l*., in 1864 by the award from the Treasury of his full salary of 2,000*l*. a year on his retirement, and in the same year by a parliamentary grant of 20,000*l*. (2) Matthew Davenport Hill, Q.C., late recorder of Birmingham; law reformer of note, especially in reference to dealings with the criminal class, substituting promptitude, certainty and strictness for delay, uncertainty and severity (see *Law Magazine*, July 1872); (3) Edwin Hill, superintendent of the stamp department; first inventor of the envelope folding-machine, since improved by Mr. De la Rue. He completely remodelled the stamping machinery at Somerset House; was most highly commended for these improvements in each of the first three reports of the commissioners of Inland Revenue, and again by a minute on his retirement, referring to his "eminent and exceptional service." He, like his brother, was a standard writer on dealings with criminals; also on currency. (4) Arthur, head-master of Bruce Castle school, where he fully developed the principles first laid down by his father; (5) Frederick Hill, formerly inspector of prisons, then assistant-secretary of the Post-office. A great and thorough reformer of the prisons under his observation, aiming to fit prisoners for honest life on their release. Concurrently, he contributed numerous memoirs on social improvements generally.

Third generation.—14 males and 17 females, among many of whom the family characteristics continue well marked. Thus (1) Dr. Berkeley Hill, and (*2*) Miss Emily Clark of Adelaide, Australia, are both actively engaged in work connected with pauper children.

LATROBE.—A family characterzied by its religious bent and musical and literary tastes, joined to a love of enterprise.

First generation.—Benjamin Latrobe, a convert to the Moravians, of which estimable sect he was a patriarch and a mainstay (Aikin's "History of Manchester").

Second generation.—3 males, 0 females; 2 at least of whom deserve notice :—(1) Christian Ignatius Latrobe, author of the well known collection of sacred music; (2) Benjamin Latrobe, architect and engineer in America.

Third generation.—7 males, 2 females, of whom 2 deserve especial notice :—(1) Charles Joseph Latrobe, governor of Victoria at the time of the gold discoveries; author of a once extremely popular book on Switzerland, called the "Alpenstock," which was the precursor of Murray's handbooks and more generally diffused knowledge. Many others of this generation, who bear the Latrobe name, are gifted with the family characteristics. (2) John Frederick Bateman, F.R.S., distinguished engineer.

Fourth generation—(still young)—includes Colonel Osman Latrobe, who was chief of General Lee's staff in America at an early age.

PLAYFAIR.—Among the characteristics of this

family is an interest in various branches of science joined to a capacity for official work and public action.

First generation.—Rev. Dr. Playfair, principal of the university of St. Andrews, author of a work on geography.

Second generation.—4 males and 3 females, of whom 3 males deserve notice:—(1) George Playfair, M.D., chief inspector-general of hospitals in Bengal; he was the head of his profession in India, and author of various medical memoirs; (2) Colonel Sir Hugh Lyon Playfair, who on his retirement from service pursued life of incessant activity in public improvement (numerous biographical notices were written of him soon after his death); (3) Colonel William Playfair, whose memory still lives in India as one of the most accomplished amateur actors.

There were two cousins in this generation, the one a very distinguished man, Professor Playfair, the celebrated mathematician, and author of the "Huttonian Theory," the other was Mr. Playfair, an architect of much eminence to whom many of the principal public buildings in Edinburgh are due.

Third generation.—21 males and 20 females, of whom 2 males deserve especial notice:—(1) The Right Hon. Lyon Playfair, M.P., F.R.S., formerly professor of chemistry, long engaged in scientific administration of various kinds, and postmaster-general at the close of the late administration; (2) Colonel R. L. Playfair, R.A., the well-known consul-general of Algiers, and naturalist. A third brother is a professor at King's College.

ROSCOE.—The type of this family is strongly marked; it has been characterized by much cultivation, refinement, and poetical taste.

First generation.—William Roscoe, author of "Lorenzo di Medici," "Leo X." &c. The above mentioned characteristics were strongly marked in him. (Life by his son, Memoirs by Hartley Coleridge in "Northern Worthies," and "Sketches" by Washington Irving.)

Second generation.—7 males and 3 females, of whom 4 males and 2 females deserve notice:—

(1) Thomas Roscoe, editor of Lanzi's "History of Painting," and author of many other works; (2) Henry Roscoe, author of a standard book on the "Law of Evidence," of "British Lawyers," and of the Life of his father; (3) and (4), both decidedly gifted, and authors of poems of merit; (*1*) Jane Elizabeth Roscoe, a woman of superior mind, intensely interested in public affairs, writer of some poems; (*2*) Mary Anne Roscoe, authoress of poems of merit.

Third generation.—17 males, 16 females, of whom 3 males and 1 female deserve notice:— (1) William Caldwell Roscoe, poet and critic (memoirs and collected works by R. H. Hutton); (2) Henry Enfield Roscoe, F.R.S., professor, eminent chemist; (3) William Stanley Jevons, F.R.S., professor, author of the "Coal Question," and of various works on logic and political economy: (*1*) Margaret Roscoe, afterwards Mrs. Sandbach, novelist.

STRACHEY.—An old family, small in numbers, but of a marked and persistent type.

Among its characteristics are an active interest in public matters, and an administrative aptitude.

There have been men of eminence in generations previous to those mentioned below.

First generation.—Sir Henry Strachey, under-secretary of state, and otherwise employed in high official posts in India, America, and England; real negotiator of Peace of Versailles (Stanhope's "History of England"); received medal of Society of Arts for having introduced indigo into Florida.

Second generation.—3 males, 1 female, of whom 2 males deserve notice:—(1) Sir Henry Strachey, Indian judge, called by James Mill, in his "History of India," "the wisest of the Company's servants;" aided much in the organization of the Indian judicial administration; (2) Edward Strachey, author of reports of acknowledged weight on Indian judicial subjects (Vth Report).

Third generation.—6 males and 1 female, of whom 3 males deserve notice:—(1) Sir John Strachey, eminent in all branches of civil administration in India; (2) Henry Strachey, Thibetan explorer, gold medallist of the Royal Geographical Society; (3) Major-General Richard Strachey, R.E., F.R.S., active administrator of Indian engineering work; physical geographer.

TAYLORS OF ONGAR.—Numerous members of this family have shown a curious combination of restless literary talent, artistic taste, evangelical disposition, and mechanical aptitudes. There is an interesting work published upon it, called "The Family Pen," by the Rev. Isaac Taylor, 1867 (see below in the "fourth generation"), which contains a list of 90 publications by 10 different members of the family, up to that time; and there have been more publications, and at least one new writer, since.

First generation.—Isaac Taylor came to London with an artist's ambition, and ended by being a reputable engraver. He acted for many years as secretary to the Incorporated Society of Artists of Great Britain, which was the forerunner of the Royal Academy. All the family characteristics were strongly marked in him.

Second generation consisted of 3 males, all of whom deserve notice :—(1) Charles Taylor, a learned recluse, editor of Calmet's Bible; (2) Rev. Isaac Taylor, author of "Scenes in Europe," &c., educated as an engraver, and far surpassing his father in ability. He married Ann Martyn, a woman of reputed genius, authoress of the "Family Mansion," and the numerous able members of the Taylor family for the two next generations sprung, with one exception, from this fortunate union; (3) Josiah Taylor, eminent publisher of architectural works; he made a large fortune.

Third generation. — Descendants of Isaac Taylor and Ann Martyn, 3 males and 3 females, of whom 2 males and 2 females deserve notice :—(1) Isaac Taylor, author of "Natural History of Enthusiasm;" (2) Jeffreys Taylor, author of "Ralph Richards," "Young Islanders," &c. ; (*1*) and (*2*), Ann and Jane Taylor, joint authors of "Original Poems" (Ann married the Rev. Joseph Gilbert). In this same generation is ranked the Rev. Howard Hinton, a leading Baptist minister, who was a son of one of the sisters in the previous generation, and is father of a well-known aurist.

Fourth generation.—6 males and 9 females now living, and some few others who are deceased; of these, 5 males and 1 female deserve special notice :—(1) Rev. Isaac Taylor, author of "Words and Places," of "The Family Pen," and of "Etruscan Researches;" (2) Josiah Gilbert, author of "The Dolomite Mountains;" (3) Joseph Gilbert, F.R.S., eminent for his chemical and physiological researches in their relation to agriculture (the paternal race of Gilbert had also a marked type); (4) Thomas Martyn Herbert, Independent minister, scholar, and writer; (5) Edward Gilbert Herbert, of the Chancery bar, who died young of diphtheria; (*1*) Helen Taylor, authoress of "Sabbath Bells."

WEDGEWOOD.—This family is curious for the sporadic character of its ability, as shown by the number of its members in rather distant relationships who have become distinguished.

The Wedgewoods must originally have been of a pure type, because the name was prevalent in the village where the great potter was born, and the bearers of it were largely inter-related, and followed the same craft. He himself married a Wedgewood, who was a third cousin, and both his father and grandfather were potters. (Meteyard's "Life.")

First generation.—Josiah Wedgewood, F.R.S., "Father of British Pottery," whose once abundant works now fetch fabulous prices.

Second generation.—3 sons and 4 daughters; 1 son deserves notice, viz.: Thomas Wedgewood, who died young. His abilities were great; he was an ardent experimentalist, and has some claim to rank as the first person who ever made a photograph. (See p. 7.)

Third generation, including descendants from the sisters of Josiah Wedgewood, contains :—(1) Hensleigh Wedgewood (English Dictionary and "Origin of Language"); (2) Charles Darwin, F.R.S. (see under Darwin); (3) Sir Henry Holland, Bart., M.D., F.R.S., who died subsequently to my having begun this inquiry; (4) S. H.

Parkes, M.D., F.R.S., professor of hygiéne to the Army Medical School.

Fourth generation.—(See under Darwin.)

STATISTICAL RESULTS.

Let us now look at the near relations of the scientific men from a purely statistical point of view, combining those already quoted with the rest, and calculate the proportion of them who have achieved distinction. It appears from my returns, which are rather troublesome to deal with, owing to incompleteness of information, that 120 scientific men have certainly not more than 250 brothers, 460 uncles, and 1,200 male cousins who reach adult life. They have somewhat *less* than 120 fathers and 240 grandfathers, because the list contains brothers and cousins. I will take two groups :—(1) grandfathers and uncles, both paternal and maternal, say about 660 persons; (2) brothers and male cousins on both sides, 1,450 persons. On the supposition, which is somewhat in excess of the fact, that I am dealing with complete information concerning the families of 120 scientific men :—

I find in the first group of 660 persons :— (1) Jeremy Bentham, a great leader of thought and founder of a school of philosophy; (2) Wedgewood, the founder of a national industry and art; (3) Compton, the inventor of a machine for cotton manufacture, which gave a timely impetus to that great national industry; (4) Maskelyne, an astronomer-royal; (5) Playfair, the scientific head of a Scotch university; (6) William Smith, founder of British geology; (7) Harcourt, the lawgiver and first president of the British Association; (8) Pemberton Milnes, who refused both a secretaryship of state and a peerage; (9) Latrobe, who was to the very worthy sect of the Moravians much what Barclay was to the Quakers, that is to say, not its founder, but a great support to it; (10 and 11) two archbishops, Harcourt of York and Brodrick of Cashel; (12) Erasmus Darwin, poet and philosopher of high repute in his day; (13) Isaac Taylor, author of "Natural History of Enthusiasm," &c. I will stop here, though it would be easy to extend the list considerably, if I took a slightly lower level of celebrity for my limit.

Every one of these 13 men when he died, was, or would have been, if he had not previously outlived his reputation, the subject of numerous obituary notices, and his death an event of sufficient public interest to warrant his being reckoned as an "eminent man." I formerly calculated, and have since seen no reason to doubt my conclusions, that the annual obituary of the United Kingdom does not include more than 50 men who are eminent in that sense. Therefore this small band of 660 individuals, contains almost one-fourth as much eminence as is *annually* produced by the United Kingdom. A different criterion of eminence may be found in the number of celebrated men reared in the universities, whither a large proportion of the brightest youths of the nation find their way. I examined the list of honours at Cambridge in the ten years 1820–9 inclusive, and also the four years 1842–5, of which I happen to have some personal knowledge, whence it appeared to me that on the average, 660 Cambridge students do not produce more than 3 men whose general eminence is of equal rank to that of the 13 men in the 660 grandfathers and uncles under consideration. A more exact test, and the best of which I can think, is to examine into the fate of the boys at large schools. It is not difficult to learn the productiveness of each school as regards eminence, because there are annual gatherings, to which former schoolboys who have won distinction are generally invited and not unfrequently come. As men begin to distinguish themselves at 35, and may be supposed willing to attend on such occasions till 70, the notabilities invited to be present at school gatherings represent the product of, say, 35 years. I feel sure that 660 middle-class boys do not turn out more than a fraction of one eminent man, though they may turn out many who do well in life and earn fortunes and local repute.

The second of the groups consists as already mentioned, of brothers and male cousins, making a total of about 1,450 men. I will examine the achievements of these, solely in respect to high university success, partly because several of the cousins are too young to have had time fully to distinguish themselves otherwise. Let us limit ourselves to the following names (the list would be lengthened if we took a lower level) :—Cambridge : (1) Alderson, both first classic and senior wrangler, that is, first mathematician of his year at Cambridge; (2) Woodhouse, senior wrangler; (3) Main, senior wrangler; (4) Humphrey, senior classic; (5) Scott, joint senior classic. Oxford: here the method of examination affords no means of ascertaining who is absolutely the first of his year, since the men are grouped alphabetically in classes, and not according to their order of merit in those classes. The names I will select are those of men who were in the first class and have subsequently distinguished themselves, viz.: (6) Moberly, head master of Winchester, now Bishop of Salisbury; (7) Francis Palgrave, critic ; (8) Hon. George Brodrick, first class both in classics and history, well known as an influential though anonymous writer. It is a remarkable fact or coincidence, that 5 men out of a group of 1,450, or say 1 out of every 300, should be first of his year in the single university of Cambridge, either in mathematics or in classics. This is about the proportion that exists among the men who actually *go* to Cambridge, and these, as before mentioned, are no chance selections, but include a large part of the annual pick of the intellectual flower of the whole nation. Moreover, these distinguished brothers and cousins of scientific men are themselves inter-related; the two senior wranglers, Alderson and Woodhouse, being first cousins, and the two classics, Scott and Brodrick, being first cousins also; both families being, in other respects, rich in ability.

We may otherwise appreciate the influence of heredity, as distinguished from that of tradition and education, by observing the similarity of disposition that sometimes prevails among numerous scattered branches of the same family. The two following extracts from the replies I have received, are illustrations of what I mean :—

(1) "My numerous relatives, though unknown to fame, are mostly characterised by great breadth of thought and rare independence of action." [These characteristics seem clearly traced by the writer to a great grandparent who immigrated from Germany]; (2) "Counting third cousins, I have scores and scores of relatives, and scarcely an *unsteady* person among them."

I have numerous returns, in which the writer analyzes his own nature, and confidently ascribes different parts of it to different ancestors. One correspondent has ingeniously written out his natural characteristics in red, blue, and black inks, according to their origin—a method by which its anatomy is displayed at a glance.

My data afford an approximate estimate of the ratio, according to which effective ability (hereditary gifts *plus* education *plus* opportunity) is distributed throughout the different degrees of kinship. They state—(1) the number of kinsmen in the several near degrees; (2) the number of those among them who were in any sense public men; and (3) the number of those who, not being publicly known, had nevertheless considerable reputation among their friends. It is therefore only requisite (after some previous revision) to add the returns together, and to compare the number of distinguished kinsmen in the various degrees with the total number of kinsmen in those degrees, to obtain results whose *ratio to one another* is the one we are in search of. These conclusions are not materially vitiated by the fact that different correspondents may have different estimates of what constitutes distinction, so long as each writer is consistent to his own scale. I have tried the figures in many ways—without any revision at all, with moderate revision, and with careful sifting, and I find the proportions to come out much the same in every case. In comparing these with previous results, obtained from an analysis of men of much higher general eminence ("Hereditary Genius," p. 317), I find the falling off in ability from the central figure, the hero of the family, to be less rapid as the distance of the kinship increases. There is however one group in that book, consisting of divines, whose general eminence is not so great as the rest, and which also resembles the scientific men in the family distribution of ability. My former figures for 100 divines gave 22 notable fathers, 42 brothers, 28 grandfathers, and 42 uncles; my present results for 100 scientific men are 28, 36, 20, and 40 respectively.

As regards the relative influence of the paternal and maternal lines, I find close equality. My method of comparison is by setting off paternal grandfathers and paternal uncles against maternal grandfathers and maternal uncles, no other near degree of kinship being available for the purpose. My results for 100 scientific men are:—paternal grandfathers, public characters, 10; of high private reputation, 3; paternal uncles, 13 and 8; making a total on the paternal side of 34. On the other hand, the maternal grandfathers are 11 and 4; maternal uncles, 15 and 7; making a total on the maternal side of 37.[1]

[1] In "Hereditary Genius," p. 196, having fewer cases of scientific men to deal with, I extended my inquiries to

I leave to another chapter some remarks about the relative value of maternal and paternal educational influences on scientific men.

nephews and grandsons, and in a second table even to great-grandparents, great-grandsons, and other equally remote degrees, but this latter was confessedly of little value.

CHAPTER II.

QUALITIES.

Energy—Size of Head—Health—Perseverance — Practical Business Habits—Memory—Independence of Character —Mechanical Aptitude—Religious Bias—Truthfulness.

IN this chapter I will speak of the qualities which the returns specify as most conspicuous in scientific men, and I shall endeavour to make them tell their own tale by quoting anonymous extracts from their communications. Some of these qualities are common to all men who succeed in life, others—such as the love for science— are more or less special to scientific men. We will begin with the general qualities, with the view of obtaining as exact an idea as may be of the degree in which they are present in the leaders of science of the present day, neither exaggerating nor under-estimating.

ENERGY.

When energy, or the secretion of nervous force, is small, the powers of the man are overtasked by his daily duties, his health gives way, and he is soon weeded out of existence by the process of natural selection; when moderate, it just suffices for the duties and ordinary amusements of his life: he lives, as it were, up to his income, and has nothing to spare. When it is large, he has a surplus to get rid of, or direct, according to his tastes. It may break out in some illegitimate way, or he may utilise it, perhaps in the pursuit of science. It will be seen that the leading scientific men are generally endowed with great energy; many of the most successful among them have laboured as earnest amateurs in extra-professional hours, working far into the night. They have climbed the long and steep ascent from the lower to the upper ranks of life; they have learnt where the opportunities of learning were few; they have built up fortunes by perseverance and intelligence, and at the same time have distinguished themselves as original investigators in non-remunerative branches of science. There are other scientific men who possess what is sometimes called quiet energy; their vital engine is powerful, but the steam is rarely turned fully on. Again, there are others who have fine intellects, without much energy; but these latter classes are quite in the minority. The typical man of science has been at full work from boyhood to old age, and has exuberant spirits and love of adventure in his short holidays, when the engine of his life runs free—temporarily detached from its laborious tasks.

We must be on our guard against estimating a man's energy too strictly by the work he accomplishes, because it makes great difference whether he loves his work or not. A man with no interest is rapidly fagged. Prisoners are well nourished and cared for, but they cannot perform the task of an ill-fed and ill-housed labourer. Whenever they are forced to do more than their usual small amount they show all the symptoms of being overtasked, and sicken. An army in retreat suffers in every way, while one in the advance, being full of hope, may perform prodigious feats.

In the following extracts I insert everything that seems deserving of mention as regards the energy of either parent. It will be observed how strong is the tendency for this primary quality to be transmitted hereditarily.

Speaking generally of these and all other extracts printed in this book, I should give the following explanation:—

Whenever anything is interpolated by me it is put in square brackets [ ]. All proper names are replaced by dots, because I do not wish to administer to the love of gossip. It is indeed impossible to prevent intimate friends from sometimes guessing the name of the author, but I have taken care that nothing is inserted which can cause annoyance. I have taken some trifling editorial liberties, such as occasionally working the words of the question into the answer, when the latter was too curt to explain itself; and in a few cases the third person has been turned into the first, for the sake of uniformity.

Extracts from Returns

ENERGY MUCH ABOVE THE AVERAGE—FORTY CASES.

1. "Travelling almost continually from 1846 up to the present time. Restless. All life accustomed to extremely rough travel; often months without house or tent. *Of mind*—restless.

"*Father*—Very energetic; restless. In old age travelled considerably. Mentally restless. *Mother*—Quiet and delicate."

2. "When young, and to æt. thirty or more, worked habitually till two and three A.M., often all night. Travelled much in various climates. Much endurance of fatigue and hard living—[an excellent mountaineer]. *Of mind*—[has risen to the highest position in his branch of science and conducts an enormous correspondence on a variety of technical and scientific subjects].

"*Father*—Very considerable energy both in body and mind. *Mother*—Below the average in bodily energy, but remarkably active mentally."

3. "When fishing or shooting (my only occupation during the holidays) I am the whole day on my legs. *Of mind*—In thirteen years I examined and named some 40,000 examples, described about 7,000 species, wrote some 6,000 pages of printed matter, carrying on at the same time a great deal of correspondence.

"*Father*—I cannot say. *Mother*—Is active the whole day. At the age of sixty-three she took sole charge of my child, then but a few weeks old, nursing it for three years, night and day. Energy of mind equal to that of her body."

4. "Remarkable energy and activity of body, and power of enduring fatigue and going without food. Extremely fond of and an adept at all field sports. Abstemious. *Of mind*—Vigorous pursuit of scientific experiments and investigations, of investment and management of money, business transactions, &c.

"*Father*—Active in field sports; has ridden sixty miles before dinner. Abstemious. Energetic in mind. *Mother*—Much energy, as shown by activity and power of enduring fatigue. Great physical courage and presence of mind in danger."

5. "Remarkable for athletic exercises when at Cambridge. In early life encountered great fatigue with the army, as . . . . during the . . . . war.

"*Father*—Great activity and immense energy in the practice of his profession. A man of most powerful intellect."

6. "I have been and still am a strong walker, both mountaineering and deer-stalking. I never knew what it was to be tired, but, after the hardest day, was ready to start again with six hours' sleep. Although in my sixty-seventh year, I am still an indefatigable deer-stalker."

7. "Strong when young—walked many a time fifty miles a day without fatigue, and kept up five miles an hour for three or four hours.

"*Father*—Remarkable energy of body up to the age of thirty, as shown . . . . . *Of mind*—Remarkable energy from early youth to his death (brought on by accident at seventy-three), when he was as actively engaged as ever in preparing for experiments [official and of a very multifarious kind]. *Mother*—Remarkable energy of mind in assisting her father in the preparation of his lectures, and afterwards her husband in his official correspondence and writings. After his death she wrote largely in magazines, and æt. eighty-five published " Suggestions for . . . . [certain improvements in administration]."

8. " When under twenty, have walked twenty miles before breakfast; when about thirty-two, walked forty-five miles; dined and danced till two in the morning without fatigue. At the age of twenty-six, during fourteen days, was only three hours per night in bed, and on two of the nights was up all night preparing for . . . [certain scientific work.] Fond of mountaineering."

9. " Considerable energy and power of enduring fatigue; rough travelling on small means

in . . . [partly-civilized countries.] Have rowed myself in a skiff 105 miles in twenty-one hours whilst undergraduate at . . . ; rowed in every race during my stay at the university; rowed two years in the university crew [Oxford and Cambridge races.]

"*Father*—[Many examples of his energy in his . . . life.] *Of mind*—considerable, compiling and writing on a great variety of subjects, whilst at the same time carrying on a system of . . . observations, and for years together. *Mother*—Energy of mind very similar to that of my father; joining nightly in . . . observations, daily in writing or drawing . . ."

10 "Very active in business, preferring walking to the compulsory driving; occupied fourteen or fifteen hours a day without distress; restlessness kept under conscious restraint; longing for adventurous travel, but hindered. *Of mind*—I doubt whether anyone in my profession has done more work, if I may reckon the total work done in . . . &c., &c.; and I worked nearly as hard while a student.

"*Father*—As a young man, an active cricketer and volunteer officer. A very earnest, active man in business, heavily engaged in it from the age of eighteen. Besides, he took an active part in town affairs and the management of many associations. *Mother*—A good walker, very active in the management of her house. Although she had a very large family, and took most diligent care of them, she was always at work, collecting all manner of things, arranging, describing, corresponding, painting, copying; she was never idle."

11. "I seem to possess the same unweariedness as my father, and find myself trotting in the streets as my father used to do.

"*Father*—Was very untiring; he tells me he has ridden 100 miles in a day. He could walk up one of the North Wales hills when nearly seventy, and used to go long distances in London, passing often from a walk into a run."

12. "In early life, occasionally working the night through. Great adroitness at games; fast runner; got the prize for fencing at . . . On

board a man-of-war in 18 . . did feats of agility, such as going up a rope hand over hand, which none of the midshipmen would attempt.

"*Father*—Great amount of quiet energy. In mind, great energy and perseverance, which lasted to the end of his life. Thus he had known little Greek, but studied it when an old man for the sake of his . . . researches; also Aramaic. *Mother* —Active housemother."

13. "Habitually travel by night without interfering with work of any kind carried on during the day. Active habits and great power of enduring fatigue."

14. "I was in youth and early manhood bodily active, a good runner and leaper, excelling almost all my schoolfellows [the school was a large one] in both points, and a persistent walker. *In mind*—During the best fifty years of my life I went through a large amount of brain-work, and vigorously pursued the several interests indicated in the enumeration of my several occupations.

"*Father*—In bodily activity much like myself, with the addition that he was a good swimmer. *In mind*—Capable of great occasional exertion rather than of sustained effort. *Mother* —*In mind*, very energetic within a limited range. Always showed great courage, fortitude, and equanimity. In her nursing duties, whether of young or old, was active, persevering, and remarkably successful."

15. " At the age of sixty made a tour, chiefly pedestrian, of four weeks in the Alps ; ascended Cima di Jazi ; crossed St. Théodule Pass, walking sometimes thirty miles a day ; æt. 67, grouse-shooting and deer-stalking. Walk six miles daily to present date. *Of mind*—See list and dates of works and papers [an enormous amount of work].

"*Father*— Active disposition ; he let his family estate, entered largely into mercantile pursuits, and died [abroad]."

16. "When young, a very quick runner and jumper ; good shot with a bow and arrow. In middle age, walked to extent of twenty-five miles a day for many months, forty miles in one day,

rarely tired. *Of mind*—In early life, any amount, provided the subject was interesting."

17. "At times, great fatigue has been gone through in connection with my profession. *In mind*—A good deal of continued power of brainwork; mental fatigue is a sensation not known.

"*Father*—Very energetic. In mind, remarkably so. Having been ruined in early life, he articled himself to a solicitor when he was thirty-five years of age; procured good practice, and wrote [a small technical book] on law. *Mother*—Loved to go through much fatigue. In mind very energetic; added greatly to the income of her family by her writings."

18. "Active bodily work an absolute necessity of my being; without it my epigastrium would gnaw itself into fiddle-strings. *In mind*—My scientific works must answer this question [they are very considerable].

"*Father*—Decidedly active and energetic; used to go out fossil-hunting when it was too late to follow his occupation [which involved out-of-door work, lasting all day and fatiguing to the muscles]. *Mother*—Very industrious."

19.—" Excelled at school and college in athletic sports, especially in long jumping (18 feet). *In mind*—Almost incapable of fatigue up to the age of thirty-eight. Usually engaged in literary work until long after midnight.

"*Father*—Remarkably active habits; a great reader when not engaged in drawing and writing."

20. " Excellent walker; great endurance of fatigue [facts are given.] *In mind*—Active mental effort all my life; have had abundance of active employment; am now doing duty as . . . . [numerous honorary offices of the first rank in importance and labour.]

"*Father*—Energetic, with considerable endurance; good swimmer. *In mind*, he had much the same active employment as myself; he took an active share in science, politics, and in religion. *Mother*—Active habits; she had great power of doing work and carrying on business."

21. "When a boy of thirteen I walked forty-eight miles in one day, fifty the next, and about twenty the third; when grown up, my powers were ordinary, certainly not above the average. *In mind*—Naturally indolent; disinclined to work unless with a large object. [N.B. I insert this moderate statement because my correspondent adheres to it verbally, and gives facts and reasons which I cannot controvert; nevertheless, if energy is to be measured by work actually accomplished, and if my correspondent's work be compared with that of other men, the estimate of his energy would be prodigiously increased.]

"*Father*—When a young man he and two brothers walked sixty miles in one day. Much mental energy; ready for all purposes. When old he was astonished at the amount of work in . . . . he did when young. *Mother*—Ordinary, both bodily and mental."

22. "Has done his chief brain-work between ten p.m. and two a.m., besides all the day labour; rests perfectly during a night railway journey.

"*Father*—Great energy, and very active; capable of enduring great fatigue."

23. "Active and energetic from infancy to eighty-four years of age. *In mind*—I must leave my works to answer this question; but I believe I have been a hard worker during the whole period of my existence. [N.B. No doubt of it.]

"*Father*—Energetic, both in body and mind; muscular; a great reader. *Mother*—Delicate, but active and intelligent."

24. "A strong walker and oarsman; can write more rapidly than any man I ever met (thirty folios of seventy-two words, equal to 2160 words an hour.) *In mind*—Have always worked long hours and very fast.

"*Father*—Remarkable energy and endurance, notwithstanding asthma: very hardworking as a . . . . *Mother*—Physically weak, but has had a large family; has done a great deal of original as well as of steady work."

25. "I am a hard rider with hounds, fond of mountaineering, and not easily tired.

"*Father*—An active man all his life, riding every day, and always about, although over eighty."

26. "Energy shown by much activity, and, whilst I had health, power of resisting fatigue. I and one other man were alone able to fetch water for a large party of officers and men utterly prostrated [other facts given in illustration of undoubted energy.] *In mind*—Shown by vigorous and long-continued work on same subject, as twenty years on . . . . and nine years on . . . .

"*Father*—Great power of endurance, although feeling much fatigue, as after consultations after long journeys; very active; not restless. *In mind*—Habitually very active, as shown in conversation with a succession of people during the whole day."

27. "Considerable enduring power in fulfilling any given task or duty; have dissected continually for three or four weeks eight or nine hours a day, devoting some sixteen hours to the work at critical times. *In mind*—Considerable.

Wrote and superintended first edition of . . . . , giving instructions to artists regarding from 200 to 300 woodcuts, correcting press, &c., without assistance, in about seven months [all this in addition to professional work]; hard work for mind as well as body."

28. " Energetic. *In mind*—[extraordinarily so, both in administrative and in original work].

"*Father*—Energetic. Author of, I think, more than seventy scientific memoirs."

29. [Formerly great power of railway travel without fatigue. *In mind*—Active and energetic in a very high degree, as shown by the amount of his official and private work].

"*Father*—Always on horseback; travelled very constantly and rapidly. Steady in pursuit of an object. He would break in horses with great skill and patience; would learn languages with great perseverance, even after fifty years of age. *Mother*—Very energetic in . . . inquiries."

30. " Great activity at cricket and football up to age of twenty-five. Captain of . . . . eleven

for five years; used to row a great deal in heavy boats."

31. " I possess considerable bodily energy, and when young excelled in fencing, swimming, and the high jump. *In mind*—Have worked hard with my brain for the last thirty-five years, almost without intermission.

" *Father*—Considerable bodily energy, and a good pedestrian. *Mother*—Sluggish bodily powers, but in mind most energetic when once roused to action by a subject that interested her feelings."

32. " Sufficiently patient of ordinary fatigue, cold, and hunger, to enable me to enjoy travelling in unfrequented countries when my companions suffered much discomfort. *In mind*—Can commonly work from twelve to fourteen hours a day without any remarkable amount of exhaustion.

Father—Capable of enduring fatigue."

33. [This is a case of extraordinary mental activity, as shown by evidence which I do not feel justified in quoting. It was rewarded by

success, notwithstanding serious impediments in boyhood].

"*Father*—A most energetic man; all for practical pursuits. *Mother*—An unusually strong mind, and steady love and search for knowledge."

34. "Walked from Cambridge to London in a day. At the age of sixty-eight ascended the Piz Corvatsch, in the Engadine. *In mind.* [Facts evidencing considerable energy are quoted.]

"*Father*—Fond of exercise; a good walker. *Mother*—decidedly active bodily habits."

35. "I am decidedly lazy; but with due stimulus could always get through a great amount of physical work, and was rather the better for it. *In mind*—As a boy, I worked for three months all day and all night, with not more than four or five hours' sleep. When full of a subject and interested in it, I have written for seven or eight hours without interruption, and without feeling any notable fatigue."

36. "In early life as a boy I was engaged in business from twelve to fourteen hours a day, yet

always found time to study and make my own instruments. Later on, my studies and scientific work were always accomplished after business hours; and it was generally my habit to commence work after dinner, and to work in science until two, three, or four in the morning, and to begin work in business again at nine. I never thought of rest if I had anything in hand of interest.

" *Father*—Remarkably active and capable of sustaining an amount of bodily exertion which would have destroyed the health of most men; for example, I have known him sustain great fatigue for eighteen hours out of the twenty-four for months at a stretch. A great walker. *In mind*—Of indomitable activity; a great reader; always at work in applying discoveries in . . . . to the arts; an untiring worker in anything he undertook. *Mother*—Busily active; great and rapid reader of current literature—perhaps had read almost every book of interest in fiction which appeared."

37. "Used to work all day at business, and one half or three-quarters of the night at science.

From Saturday afternoons to Monday mornings would walk forty to fifty miles [in pursuit of a branch of natural history]. Could work hard at business all day (and a very anxious business), and at evening and night would work hard at [two branches of science]. Found a wonderful relief in science.

"*Father*—Energetic in travelling; great energy in business."

38. "For several years was engaged in full medical practice, and at the same time was a lecturer on . . . and engaged in investigations on . . . . for which the Royal medal was awarded by the Royal Society.

"*Father* and *Mother*—Both of active habits."

39. "In professional life I have often been up three successive nights without distress, but did not like a fourth, if it came. Consider *that* my limit. *In mind*—Wrote . . . . [a considerable work] between eleven P.M. and two A.M., after professional hours. All the time that I have devoted to science has been stolen from strictly

professional engagements, but more often from myself."

40. "Considerable power in earlier days of enduring mental fatigue and of taking up without difficulty a considerable range of subjects. Example:—I was for a little while, æt. seventeen to twenty, employed in teaching, and I contrived in my scanty intervals of leisure to read a very large quantity of Greek and Latin, and to become, without any external assistance, a very fair mathematician [my correspondent occupies a high official position, in which considerable mathematical knowledge is essential]. I learnt also Italian at this time."

41. "I should say considerable, judging by the number of things I have been able to learn and to do since adult age."

42. "I think considerable, in mind. Have commonly had it said of me that it was wonderful how I got through so much work.
Father—Was well known as a hard-worker. *Mother*—A great reader; taught herself Greek

and Hebrew, and learnt German, in later life, to read Luther and other theological writers in the original. A great student of theology."

CASES OF ENERGY BELOW THE AVERAGE—TWO CASES.

1. "No remarkable energy of body. *In mind*—Never capable of a large *amount* of brain-work; for years have regarded myself as defective in brain-power. [The actual performance of this correspondent is considerable, and of a very high order.]

"*Father*—In early life fond of athletic sports, and an enthusiastic sportsman. Energy of mind very remarkable, shown in early university and professional life and all subsequent occupations. He wrote a large number of publications on subjects of . . . and . . . . controversy. *Mother*—Energy of mind remarkable; zeal in pursuit of interests, excessive."

2. "Constitutionally languid, with a strong wish for greater energy and more power of en-

during fatigue. *In mind*—Energetic as far as health permits. Much occupied professionally, but when well, capable of vigorously following up the science of . . . . in leisure hours.

"*Father*—Energetic in body as far as his health allowed; in mind, very energetic. His brain-work from an early age was very large in amount, and he was vigorous and sanguine about anything he undertook. *Mother*—Very languid; incapable of any bodily exertion. Very little energy of mind; too languid to take much interest in anything beyond her own family."

SIZE OF HEAD.

I may mention that energy appears to be correlated with smallness of head, a fact which is well illustrated here, although the average circumference of head among the scientific men is great. Energy is also, as we have seen, strongly marked among them; but it is much more strongly marked among those who have small heads. I have ninety-nine returns, many of which I have verified myself, using the hat maker's whalebone hoop, and measuring inside the hats. It appears that the average circumference of an English gentleman's head is $22\frac{1}{4}$ to $22\frac{1}{2}$ inches. Now, I have only thirteen cases under 22 inches, but eight cases of 24 inches or upwards. The general scientific position of the small-headed (who are mostly slender, but not necessarily short) and large-headed men seems equally good; but the fact is conspicuous that, out of the thirteen of the former, there are only two or three who have not remarkable energy; and out of the eight of the latter there is only one who has. A combination of great energy and great intellectual capacity is the most effective of all conditions; but, like the combination of swiftness and strength in muscular powers, it is very rare.

HEALTH.

The excellence of the health of the men in my list is remarkable, considering that the majority are of middle and many of advanced ages. One quarter of them state that they have excellent or very good health, a second quarter have good or fair, a third have had good health since they attained manhood, and only one quarter make complaints or reservations. Here are two examples of excellent health in which some details are given :—1. "Only absent from professional duties two days in thirty years; only two headaches in my life." The next is from a correspondent who is between 70 and 80 years of age. 2. "Never ill for more than two or three days except with neuralgia; no surgical operations except inoculation, drawing of one tooth, and cutting of corns."[1]

I may add a characteristic biographical extract from the *Times,* Oct. 31, 1873, relating to the late Sir Henry Holland, who was on my list :— "Certain it is, as all who have fallen in with him by sea or land will attest, that he might be seen in all climates, in the Arctic Regions or the Tropics, on the Prairies or the Pyramids,

---

[1] I read this at my lecture at the Royal Institution. It was from the pen of the geologist, Professor Phillips; a few days afterwards, he was killed by a fall down stairs at Oxford!

in precisely the same attire—the black dress coat in which he hurried from house to house in Mayfair. Yet he never had a serious illness till his last. There was not a day, probably not an hour, when he could not boast of the *mens sana in corpore sano;* and, without headache or heartache, he attained the extraordinary age of 86."

It is positively startling to observe in these returns the strongly hereditary character of good and indifferent constitutions. I have classified the entries, each entry giving the health of the scientific man, of his father and of his mother respectively, and find as follows :—First, a long row of such terms as these: " Excellent ; excellent; excellent ; " or " Good ; good ; good ; " then comes another row in which some ailment is specified by the scientific man as affecting himself, and as having also affected one or other of his parents. Examples :—1. "Excellent, but hay fever ; father, excellent, but severe hay fever." 2. "Good in early life, subject to headache ; father, good, subject to headache." 3. " Delicate in early life, one lung seriously affected; mother delicate and phthisical." I can find only two cases, neither very strongly marked, in which both parents are described as unhealthy, although marriages between such persons are not infrequent. The returns seem to show that the issue of these marriages are barely capable of pushing their way to the front ranks of life. All statistical data concur in proving that healthy persons are far more likely than others to have healthy progeny; and this truth cannot be too often illustrated, until it has taken such hold of the popular mind, that considerations of health and energy shall be of recognized importance in questions of marriage, as much so as the probabilities of rank and fortune.

I may mention, as a fact that corroborates my belief in the exceptionally good physique of scientific men, that I find the average height of those who have sent me returns, to be half an inch above that of their fathers.

PERSEVERANCE.

Steady perseverance is a third quality on which great stress is laid; but this might have been anticipated, and it is unnecessary to quote many instances. Here are a few :—

1. "I have probably beyond the average, steadiness of determination, even when the subject is distasteful." 2. "Steadiness decidedly marked." 3. "Determination never to leave unaccomplished a matter once taken in hand." 4. "Great continuity and steadiness." 5. "Steady and intense perseverance." 6. "Very persevering, not discouraged by defeat." 7. "Determination to succeed when possible; my motto being 'Whatever thy hand findeth to do, do it with all thy might,' for 'the night soon cometh when no man can work.'" 8. "I do all things at a white heat, but never tire of the pursuit." 9. "Continuous pursuit of certain studies from an early age." 10. "Steadiness and perseverance in the pursuit of an object is my most distinctly-marked peculiarity." 11. "The most prominent are perseverance and

industry. A willing mind and determination to persevere is, in my opinion, the most direct road to success; we must, however, exercise a sound judgment in the selection of subjects on which to exercise our thoughts."

I do not think it necessary to quote the instances where either parent is also spoken of as being remarkably persevering; these may be taken for granted. I find that the father is referred to in strong terms eight times, and the mother only twice.

As a set-off to the above, Impulsiveness is not confessed to by a single physicist, chemist, or mechanician. It is equally absent in their parents, with the exception of the mother of one of them. Among the remaining men of science, I only find 5 cases, but these are mostly combined with some tenacity of purpose, and they are all inherited.

PRACTICAL BUSINESS HABITS.

Some prevalence of practical business habits might also have been anticipated, but they prove to be much more common than I had expected. Among those who have sent me returns, I count no less than seventeen who are active heads of great commercial undertakings. There are also ten medical men in the highest rank of practice, and eighteen others who fill or have filled important official posts. Here are some answers to my special inquiries:—

1.— A most eminent biologist wrote as follows, in reply to the inquiry whether he had any special tastes bearing on scientific success, in addition to those for his own line of investigation:—" I have no special talent except for business, as evinced by keeping accounts, being regular in correspondence, and investing money very well." It is clear that method and order are essential to the man who hopes to deal successfully with masses of details.

2.—" I believe I may say that my organ of order is highly developed. Of my collection of some 7,000 birds' skins every one is always in its place, ticketed with name, &c., all by my own hand. I spend much time, perhaps too much, in putting things straight."

3.—" I believe I am reckoned a good chairman at public meetings, and I always find that administrative and other work gravitates towards my hands."

4.—" My professional life is strictly methodical; every working day is still mapped out into hours, half hours, and quarters."

Fully one half of those who state that they possess business habits in a decided degree accredit one or both of their parents with the same faculty.

Only two of my correspondents speak of being deficient in business capacities. Both these are physicists.

The following quotation may with propriety be inserted here, although the first named quality, independence, is the subject of a future chapter. "I attribute all the knowledge I have acquired, and any success I may have had, chiefly to three qualities, all of which I believe I inherited. First, independence of judgment which prompted me to learn for myself what I wanted to know. Secondly, earnestness, determination, and perseverance in acquiring such knowledge, often under difficulties, and in the face of routine business occupation; and thirdly, a business-like, practical, logical way of looking at things, which enabled me to direct attention to the important and relevant, neglecting the unimportant and irrelevant points in what I had to study and do."

MEMORY.

Memory is very variable in power and character, perhaps no other quality is more so. It is an important ingredient in that aggregate of faculties which form general scientific ability, as is shown by the fact that about one quarter of the men on my list possess it in a high degree, but it is not an essential one, because it is defective in about one case in fourteen. A good memory is of greater importance to the young student who has much to learn, than to the advanced philosopher who has chiefly to reflect, and who knows where to refer for information. Memory is usually defective in persons of small ability, but not invariably so; even among idiots it may be sharp. There are two cases of this recorded in the autobiography of the late Mrs. Somerville (p. 92.) One cannot but suspect some exaggeration in the statements, and feel regret that the cases were not fully inquired into, both as regards the precise power of memory, and the degree of development of the other faculties. She says of the first idiot, " He never failed to go to kirk, and on returning home could repeat the sermon word for word, saying, " Here the minister coughed, here he stopped to blow his nose." She then speaks of " another idiot who knew the Bible so perfectly, that if you asked him where such a verse was to be found, he could tell without hesitation and repeat the chapter."

I have sorted such of the replies as are of interest, into the following groups. (1) Good verbal memory, as for prose and poetry, 6 cases;

(2) good memory for facts and figures, 9 cases; (3) good memory for form, 6 cases; (4) good memory for names in natural history, 4 cases; (5) good memory, no details, 5 cases; (6) fitful and peculiar memory, 6 cases; (7) bad memory, 7 cases. Total number of noteworthy cases, 43. I have not included in the above, a few instances in which the scientific man has described his own memory, simply as "good," nor others in which he has made no remark, except that one of his parents had very good memory. The hereditary character of this quality is abundantly illustrated.

Good verbal memory, as for prose and poetry.

1. "Very great, both for facts and words; I could in my earlier days often retain poetry after two perusals, and once learned, it was seldom forgotten. I have seldom met a quicker or more retentive memory in any one."

2. "After reading over a lecture or speech of an hour's duration, three times, can recollect nearly the words as written for 8 or 10 days."

[I am informed verbally by this correspondent, that he is obliged to abstain from writing out his addresses, &c., beforehand, otherwise he has found the memory of what he wrote to be so strong and exacting as to make it difficult to him to deviate from it and accommodate his language to the current temper of his audience.]

"*Mother*—Excellent memory."

3. "Considerable, both verbal and objective; great facility in quotations; familiarity with large collections of coins and specimens.

"*Father* and *Mother*—both good memories."

4. "In childhood, all the Psalms, old version; much old English poetry; afterwards, nearly the whole Latin grammar (Eton), Virgil, Ovid, Lucan; still later, considerable parts of the Iliad, Odyssey, &c., could be, and partly can [still] be, repeated *ex memoriâ;* zoological, botanical, mineralogical and paleontological names in abundance."

5. "My memory was very good. I remember as a boy, to have read Schiller's 'Thirty Years'

War;' I could afterwards without effort, say pages of the work by heart."

6. "At school I used to learn in a single evening 100 lines of Virgil, and repeat them correctly in the morning.

"*Father*—very good."

Good memory for facts and figures.

1. "Next to no verbal memory, but good for facts small or great which will fit into any chain of reasoning."

2. "Of moderate verbal memory, but strongly retentive of facts and figures so far as they are related to any subject on or in which I was engaged.

"*Father*—Memory very retentive, but not systematic. He had a great amount of information, but had not great acquirements; his familiarity with Scripture was, however, remarkable. *Mother*—Very retentive for small facts and figures."

3. "My memory of things learnt early in life (as dates, rules, examples of grammar, &c.) very retentive, but of all isolated facts of subsequent occurrence, as the birthdays of my children, and the dates of events of my own life, I am singularly destitute of retentive power. On the other hand, of whatever is linked by rational association with any subject in which I take an interest, my memory is very good.

"*Father*—The power of his memory was shown by the great range of his acquirements; he had greater power of remembering isolated facts than I have."

4. "I should say far above the average. I can now refer to note-books of 30 years past and select a special observation. In other words, it is a capital working memory. I never tried to learn pages of poetry, &c.; in this I should probably have failed."

5. "Memory exceedingly strong and retentive, especially of dates, figures and events.

"*Father* and *Mother*—both had good memories."

"6 Great memory for figures; can get up pages for examination before committees, and dismiss them from memory afterwards. Strong recollection of scenery."

7. "Very retentive memory, especially of acts, circumstances, and individuals."

8. "Never kept a diary; clear remembrance of events in childhood with their dates in every year from the age of six onwards. Solve problems better out of doors than in the study. Can forget useless knowledge such as formulæ, rules, gossip, &c., very fast."

9. "Bad memory for names and dates, but good as regards facts or circumstances; principles in physical science are clearly retained.

"*Father*—Excellent memory for historical events, including dates and names in ancient and modern history. *Mother* — Moderately good."

Good memory for form.

1. "Memory most treacherous except in certain respects. Vivid and generally very accurate as

to places and visual images. As to thousands and perhaps tens of thousands of specimens and plants, can remember the exact spot where each was gathered. As to a multitude of facts that should have interested me, my memory is a blank and the original impression revived with difficulty if at all. . . . Very retentive and accurate as to the *sequence* of impressions from early childhood onwards.

"*Father* — Remarkably retentive memory; quoted long passages from classical authors not seen for a very long time previous. Shortly before his death, at 73, recited a long passage from 'Gibbon,' not read for fifty years before. *Mother*—Memory not reliable generally, but clinging strongly to special scenes and events."

2. "I recognize most of the animal forms which I have previously examined, but I forget easily the details of their structure, also their systematic names (specific, not generic). Likewise I have a good memory for faces, but not for names of persons; could never remember historical dates."

3. "Great power of remembering forms and points of objective interest; none of numbers or abstract arguments. Languages, poetry, &c., soon lost if not kept up."

4. "Strong local memory especially of scenery."

5. "Very good memory for ideas and general notions, also of persons and places seen; verbal memory not at all good. *Mother*— Good memory."

6. "Great memory for faces and objects once seen."

7. "A good memory for faces, for locality, for things, for events, for scientific facts; but not particularly good for figures or quantities, except in all necessary routine, as in prescribing and in subjects of lecture. Never failed to recall what I desired, in my lectures.

"*Father*—An excellent memory; was a very first-rate whist player. *Mother*—An excellent memory; played a capital game at whist."

Good memory for names in natural history.

The power of recollecting a multitude of grotesque and barbarous names, which all naturalists must possess to a considerable degree, and which seems so extraordinary to persons who are not naturalists, is hardly alluded to in these returns. It would appear that our most eminent naturalists are not very specially gifted among their fellow-workers in this respect. Here are a few cases of a rather good memory of the kind :—

1. "Memory strong up to the age of 38; still good and capable of recognizing and naming probably between two and three thousand species of animals and plants, including fossil forms.

"*Father*—Remarkable; capable of accurately repeating from memory the substance of speeches delivered at clerical and other meetings."

2. "Retentive of botanical names; rather deficient in other respects, especially as to persons."

3. "Retentive for nomenclature, but not for numbers or history."

4. ". . . . during practitional life I have gone over the foraminiferæ and remember all their names."

Good memory, no particulars given.

1. "Very remarkable retentiveness of memory.
"*Father*—Good. *Mother*—Very good, full of anecdote."

2. "Very good memory as far as my 85th birthday."

3. "Very good.
"*Father*—Good."

4. "Very retentive, but not exactly accurate."

5. "Retentive memory for what was of interest, and very accurate.
"*Father*—Retentive."

6. "Very good as a boy and young man."

Fitful and peculiar memory.

1. "Occasionally remarkable, but very fitful. I have occasionally been able to repeat pages after once or twice reading; at other times it is below the average. A power of eliminating and retaining the salient points of what I read, if it interests me, but very bad memory for facts and details."

2. "Although I can speak for an hour or two from a few notes, I could not repeat correctly a few sentences from memory.

"*Father*—Remarkable for good verbal memory; could repeat pages of poetry and speeches, without mistake, a striking contrast to my own memory."

3. "My father and myself have memories of the same character; treacherous in matters of business and very retentive of scraps of verse read over and learnt long ago. When my father was to have met me, a little boy returning from school at the end of the half, he would forget

all about it. My engagements sometimes suffer . . . [from similar forgetfulness]."

4. " Memory very retentive in regard to incidents and events, but could never learn by rote except with great effort. Often surprise my patients by recollection of their symptoms, but am often at a loss to connect their names with their faces.

"*Father*—Memory remarkably retentive, especially as to the various events of his life and time."

5. " Memory very bad for dates and for learning by rote, but [extraordinarily] good in retaining a general or vague recollection of many facts.

" *Father*—Wonderful memory for dates; in old age he told a person, reading aloud to him a book only once read in youth, the passages which were coming; he knew the birthdays and those of the deaths, &c., of all his friends and acquaintances."

6. " A peculiar memory; bad for names of

persons, plants, places, &c. ; good for subjects connected with others; not bad for numbers.

"*Father* — A most marvellously retentive memory; he could relate minute details of historical occurrences, names of actors in politics, almost all he had ever read (he was a great reader), and was in consequence a most lively companion. *Mother*—Not very good."

Bad memory.

1. [A physicist informs me that his memory is unable to retain even the commonest constants in habitual use, and that the selection of his special line of investigation was governed by his sense of this disability.]

2. "Bad memory; from boyhood incapable of learning school tasks by heart, though retaining a knowledge of principles and methods."

3. "I have a very poor memory; I was once a whole fortnight in recovering the name of . . . . , but I got it at last. I consider that all attempts at making me learn poetry, and in

particular Latin poetry [at school] were gross mistakes; I was never benefited in the least. Reasoning was my forte, and I could never do anything by rote."

4. "A bad memory, especially for names."

5. "Not possessed of a retentive memory either in small matters or large ones, except in those in which I take a *special* interest."

6. "I was always slow at learning."

7. "Memory not retentive; very much under the influence of association and suggestion.

"*Father*—Memory very retentive as to principles, facts, and incidents; not much so as to names of persons and objects. *Mother*—Not retentive."

INDEPENDENCE OF CHARACTER.

We now come to the qualities that are of especial service to scientific men; those already mentioned, of energy, health, steadiness of pursuit, business habits and memory, being of general utility. The first of these is independence of character. Fifty of my correspondents show that they possess it in excess, and in only two is it below par. Here are a few examples :—

1. " Left, æt. 12 [that is, ran away from], a school where I had received injustice from the master." 2. " Opinions in almost all respects opposed to those in which I was educated." 3. " I have always taken my own independent line. My heresy prevented my advancement." 4. " Preference for whatever is not the fashion, not popular, not rich, not very able to help itself, yet with qualities unworthily overlooked or unjustly oppressed."

The home atmosphere which the scientific men breathed in their youth was generally saturated with the spirit of independence. Examples :—

1. " My father was extremely independent, in some respects more so than I am. He never altered the fashion of his dress; he never took off his hat to anyone in his life, and never addressed anyone as Esq." 2. " My father was a Liberal when Liberalism (then styled Jacobinism) was highly obnoxious, an early denouncer of slavery

and advocate of religious liberty, a free-trader when the world was protectionist, and an opponent of unrighteous war when war was most popular. He was for mitigating our criminal code when hanging was regarded as the sheet-anchor, and, in a word, was politically and socially a very independent spirit." 3. "My father [an exceedingly humane and courageous man, who was a master in the Royal Navy] would never, unless compelled, attend the flogging of seamen, a punishment mercilessly and unsparingly administered in his days .(1800–1815)." 4. "It was marked in my father; he held Jacobite opinions, when it was not very safe to hold them." 5. "Maintenance by my father of religious and political creeds at a time when these creeds were unpopular and often disqualifying."

In confirmation of the assertion that the scientific men were usually brought up in families characterized by independence of disposition, I would refer to the strange variety of small and unfashionable religious sects to which they or their parents belonged. We all know that

Dalton, the discoverer of the atomic theory, and Dr. Young, of the undulatory theory of light, were both Quakers, and that Faraday was a Sandemanian. So I find in these returns numerous cases of Quaker pedigree; and I know of one man, not as yet technically on my list, who was born a Sandemanian. There are also representatives of several other small sects, as Moravians and Bible Christians, and the Unitarians are numerous. It will be understood that the object of saying this is not to throw light on the religious tendencies of the scientific men (concerning which I shall have almost immediately to speak), because so off-hand a statement would mislead, but to prove that they and their parents had the habit of doing what they preferred, without considering the fashion of the day. The man of science is thoroughly independent in character.

MECHANICAL APTITUDE.

There is a prevalent taste for mechanics among scientific men, whose peculiarity it is to be interested in things more than in persons. One would have expected to find it developed among physicists ; and, as a fact, eight of them possess it in a high degree, and similarly among mechanicians and engineers, all of whom must possess it, and four of whom testify to it, but it seems just as strong among the rest. Here are instances and extracts :—

Chemistry.—1. " Constructed a reflecting telescope, with 12-inch aperture." 2. " Ground, polished, and silvered a 7-inch glass speculum, and mounted it equatorially." *Geology.*—3. " Considerable mechanical skill." *Biology.*—4. " Always fond of constructing ; school nickname, 'Archimedes.' If I had followed my profession should probably have been [very successful as] an engineer." 5. " Very fond of mechanical contrivances. Invented and made my own toys as a child. Mechanical tastes are still largely indulged in intervals of leisure." 6. " Special love of mechanics ; a good amateur cabinet-maker and blacksmith. Made lithotrites." 7. " Talent for mechanics." 8. [Was extremely ingenious in devising modes of preserving and exhibiting objects of natural history]. 9. "Strong natural inclination towards mechanism." [His present profession was accidental and against the grain]. 10 and 11. "Aptitude for mechanism." 12. "A decided turn for mechanical pursuits, both in arrangement and construction." *Statistics.*—13. "Fond of and quick in understanding machinery." 14. "I always took great interest in mechanical improvement." 15. "I often feel a positive pain in passing an object of which I do not comprehend the meaning and construction."

RELIGIOUS BIAS.

It appears that out of every ten scientific men, seven call themselves members of the established Churches of England, Scotland, or of the now disestablished Church of Ireland, and three belong to one or more of the following sects, which I name in the order in which they are most numerously represented :—

1. None whatever; 2. established Church with qualification; 3. Unitarian; 4. Noncon-

formist; 5. Wesleyan; 6. Catholic; 7. Bible Christian. There is much Quaker, and even some Moravian blood, but there are none who have sent me returns who still profess those creeds. The creeds of the parents are somewhat more varied than the above, and the Unitarian element is stronger.

The religious feeling of men of science is necessarily of a peculiar character. Being thoughtful men, they are probably more occupied with religious ideas than the generality of people; but, being exacting of evidence and questioners of authority, they sturdily object to much that others accept easily. But what is "religion?" It is one of the vaguest of words. Let us try to express ourselves more clearly. I think we may assume that the general tendency of scientific men is to take a "philosophic" view of life; that is, to show some disregard of the petty, transient events which chiefly absorb the attention of mean minds, and to feel most at peace when their thoughts are reposing on the larger and more enduring aspects of the moral and material world. Also, it would be easy to show that no class in the community are more active as philanthropists than scientific men. But these tendencies do not cover the meaning of the phrase, "religious bias" in its technical sense. So far as I understand that sense, it comprises three elements :—

1. Great prevalence of the intuitive sentiments ; so much so, that conflicting matters of observation are apt to be laid aside, out of sight and mind. The intuitive sense of a supreme God, who communes with our hearts and directs us. 2. A sense of extreme sin and weakness, as expressed by the liturgical phrases, "No power of ourselves to help ourselves," "Through the weakness of our mortal nature we can do no good thing without Thee," &c. 3. Revelation of a future life and of other matters variously interpreted by different sects, which, more or less, satisfy the intuitive sentiments.

I did not enter into these details in framing my questions, but simply asked in general terms whether or no my correspondents had a strong religious bias. The interpretation I put on the answers which are subjoined, is that religion, in the sense of the third paragraph, is not actively accepted by many of those who describe themselves as religiously inclined : they seem singularly careless of dogma, and exempt from mysterious terror. Also, considering the independence of their disposition, their energetic temperament and healthful physique, I should think that religion, in the sense of the second paragraph—that of feeling sinful and weak—would not express the views of many of them. Therefore I look on the intuitive sentiments, as described in the first paragraph, connected with a philosophic frame of mind and a tendency to active philanthropy, as the most likely meaning of the phrase " religious bias," when it is used without any qualification by my correspondents, especially by those who are Unitarians. In this sense, at least, there appear to be about eighteen instances of scientific men who have a decided religious bias ; being, I should estimate, at the rate of two or more, in every ten ; but I am not able to state with certainty how many of these are religious in the sense of all the three paragraphs.

Religious sentiments weak, accompanied with more or less Scepticism.—1. [Being compelled to attend frequent chapels at college, he, for ten years afterwards, refused to enter either church or chapel]. 2. " The negative tendencies of my family may be absence of piety . . . ." 3. " Religious feeling not great." 4. " Sceptical." 5. " Not much religious bias except in a boundless admiration of nature." 6. " I gave up common religious belief, almost independently from my own reflection." 7. " Bias towards freedom of thought in religious matters."

Intellectual interest in religious topics.—1. " Entertained at an early age independent views regarding the resurrection and salvation of the heathen, which led to frequent disputes." 2. " At school I became a sceptic, and even worked out in my own somewhat (at that time) reserved mind, a kind of idealism. I afterwards had a phase of religious fervour, but worked through it." 3. " Given to theological ideas, and not reticent about them." 4. " Instinctive (or original) religious bias, though this may be in part due to early training. . . . I take considerable pains in the investigation of religious matters, one of my amusements being the collection of a considerable theological library, with the books of which I am familiar."

Dogmatic interest.—" I have no more doubt about the plenary inspiration of Scripture than I have about the simplest axiom in mathematics." [I class this exceptional reply under " dogmatic interest " because the remainder of the writer's brief communication hardly suggests the dependent frame of mind that is characteristic of " piety "—*e.g.*, " Never received or asked a single favour or a single farthing for anything I ever wrote or did."]

Religious bias.—1. " Religious bias." 2. " Of a religious bias of thought." 3. " Religious views liberal, but strongly anti-materialistic." 4. " Early religious impressions strong, but have, on the *dogmatic* side, quite disappeared. The belief in a permanent antithesis between good and evil, irrespective of utilitarian results, has survived, with no keen sense of the need of a dogmatic basis for the belief." 5. "Much religious bias of thought from early education." 6. " I have been the more biased towards religion, in that my father and maternal grandfather lived it and did not prate about it. I am personally only a combination of these two men in this respect (. . . Please take the *sense* of what I have written, and not the *words*)." 7. "Religious bias of thought decided." 8. "Although firmly and thoroughly believing in Christianity, and accepting it as the guide of my life, as far as I can understand it, being also a regular attendant of the Church of England, still I cannot admit the right of that or any other Church to teach dogmatically what truths are necessary for my salvation; and the feelings which ever cause me to resent any interference with the liberty of conscience are quite as strong in me as they were in the breast of my ancestor, when he gave up the land of his birth and property, more than 300 years ago." [My correspondent has shown marked instances of independence of character, and is descended maternally from both Flemish and French religious refugees, and paternally from an English Nonconformist, who left his country and settled in America.] 9. "It is difficult to estimate one's own peculiarities, but I believe I may credit myself with more than the usual amount of (. . . and) religious bias of thought. I have mixed and worked with Christians of most of the Protestant Churches." 10. "Strong religious feeling. My intention on entering . . . was to devote myself to a missionary life in China; but my unexpected success in . . . persuaded my friends to induce me to abandon my purpose, on the ground that I might serve God better in my new sphere at home. I yielded to their arguments with great reluctance." 11. "Intensely religious; formerly in the Evangelical sense (a tract distributor, promoter of prayer-meetings, bible classes, &c.) Excessive distaste to novels and fictions in any shape." (See "*Indifference to dogma*," p. 137.) 12. "I was brought up an ordinary member of the . . . Church, but ultimately came to the conclusion that . . . was essentially illogical. . . . I had the happiness of seeing my mother follow me into the . . . Church." [I regret that I am unable, with propriety, to give fuller extracts from the most interesting and instructive replies of this correspondent.]

Religious bias, with intellectual scepticism.— 1. "I have not cultivated independence of judgment in religious matters; I have shrunk from so doing in order to retain peace of mind and leisure for my varied studies." 2. "Much religious bias of thought, but no respect for revealed religion as a base for such a bias." 3. "Religious bias towards natural theology strong, as distinguished from dogma of any kind." 4. "I have, perhaps, a religious bias from early habits and associations, rather than from temperament; but I have always had more pleasure in sacred than in secular music, which, perhaps, shows the predominance of the emotional tendency." 5. "A profound religious tendency, capable of fanaticism, but tempered by no less profound theological scepticism."

Next, as regards the effect of dogmatic teaching, or of "creed," on research. I had expected it to have been much more deterrent and hindering than the answers warrant. The suicide of the geologist, Hugh Miller, whose brain gave way under the conflict between dogmatic creed and scientific doubt, is a terrible tale. One would have thought that the anathemas from the pulpits against most new scientific discoveries, as soon as they became capable of popular application, such as geological history, antiquity of man, and Darwinism, must have deterred many; and, as I have already shown, few of the sons of clergymen are on my list. Nevertheless, in answer to my direct inquiry "Has the religious creed taught in your youth had a deterrent effect on the freedom of your researches?" I am met with an overpowering majority of negatives. Seven or eight say "no," justifying their assertion by various reasons, to one who says "yes," as is shown by the appended replies. These may be sorted into the four following groups:—

(1) "No" deterrent effect—39 cases. (2)

"None," with emphasis—12 cases. Examples:
—"None whatever;" "not in the least;" "not in the slightest;" "decidedly not;" "certainly not." (3) "None," with various classes of reasons why it had not—14 cases. (4) Has had a good and not a bad effect—8 cases.

Further specimens of the first two groups "no," with or without emphasis, are needless; but I will give extracts from the remainder, divided under convenient heads.

Have no dread of inquiry.—1. "I do not think so. At the time when I held strongly the . . . dogmatic system I never could apprehend any dread of the results of free inquiry." 2. "None whatever. Absolute and fearless faith in the truth." 3. "I was left free to choose my own religion, and believe that there is no real antagonism between revealed religion and the study of nature."

Religion and science have different spheres.—1. "No; it [religious creed] has no point of contact with chemistry."

Indifference to dogma.—1. "Not in the slightest degree; but the method of science has taught me not to put any confidence in creeds or dogmatic statements of any kind." [This correspondent is the tract distributor, &c., of (11) of those having "religious bias" in p. 133.]

Liberality of early religious teaching.—1. "None. The teaching was not severe or exclusive in any degree; it was the ordinary teaching of the Church of England." 2. "My religious creed from infancy was that of freedom. I was not taught creed or dogma, and had therefore the great advantage of not having to fight my way out of darkness into light." 3. "I learnt no creed in my youth." 4. "I had no religious instruction at school." 5. "No; freedom of thought was always made a part of the creed practically taught me." 6. "No religious creed was ever taught to me." 7. "None whatever. In fact, no *creed* was taught me." 8. "My religious freedom has enabled me to look every scientific question fairly in the face." 9. "There was no religious coercive education at home.

notwithstanding my mother's strong personal religious bent. On the contrary, her influence was quite in the direction of free inquiry, in which she largely indulged herself. My school religious teaching had no effect that I can perceive, either on my intellect or imagination. Its chief result was to make me detest the drudgery of learning catechisms and sitting through dreary sermons."

[2, 3, 6, 7, 8, are children of Unitarian parents.]

Have early abandoned creeds.—1. "At æt. 13, I disbelieved as thoroughly as I do now in the religious creed (that of the Church of England) in which I was brought up; and I had realised Berkeleyan idealism by my own road." [Compare this with the reply, 2, from a different correspondent in p. 130 in the section, "*Intellectual interest in religious topics.*"] 2. "None whatever; I have long since wholly rejected religious creeds." 3. "I gave up common religious belief almost independently from my own reflection." [This quotation is repeated from the last section. The writer's reply to the question of which we are now speaking was a simple "no," and has been classified as such.]

The religious creed has had a good effect on freedom of research.—1. "None [*i.e.* no deterrent effect]; rather the contrary." 2. "On the contrary." 3. "Quite the reverse." 4. "I think none whatever. I have had to overcome some prejudices, but my true religious life has been cognate with my scientific one, and the former has stimulated rather than crippled the latter." 5. "Certainly not! On the contrary, it has had clearly the very best effect." 6. "Not a deterrent effect; but it acted as a guide." 7. "Never deterred; now acts as a direct stimulant, since it appears to me that the cultivation of a naturally-implanted intellectual tendency is a religious duty. . . . The most pernicious influence to which I was subjected was that arising from J. Stuart Mill. It took me a long time to work through the sensationalist, empirical philosophy, and to come out at the other side." 8. "No; but the scientific system inculcated long prevented me giving my religious feelings and aspirations full sway."

Has had some deterrent effect.—1. "Certainly the narrow . . . ism of early youth made me for a long time a timid thinker." 2. "To a certain extent, yes—not in philosophical research; but I shrink from the disturbance of mind (not fear of ultimate consequences) which I know would follow diving into certain questions of the day, connected with early religious teachings." 3. "No; for some time it may have hindered me." 4. "It certainly would have had that tendency, though not that effect, if my researches had taken certain directions." 5. "Would have been so had I not fought against it." 6. "The 'Biblical faith' prevented my getting good geological views for many years, by having set my thoughts in the old grooves, and thus limited them." 7. "I think not. I emancipated myself from dogmatic trammels early in life, but not without a struggle." 8. "After about ten years' careful consideration of the facts, called by theology 'seeming contradictions of science,' I finally discarded the pentateuchal spectacles through which I had previously looked at certain phenomena. I lay to early theological teaching so much hindrance in the quest of the most precious of our possessions—truth."

TRUTHFULNESS.

A curiosity about facts is much spoken of and implied in the answers to my questions; in a few cases it is combined with a curious repugnance to works of avowed fiction. A hunger for truth is a frequent ingredient in the disposition of the abler men of every career; but in all probability it is felt most strongly and continuously by men of science. The most clearly-marked characteristic of scientific society seems to me to lie in the careful accuracy with which facts and anecdotes of all kinds are related. I have the good fortune to be acquainted with a large family circle whose curiosity about facts and practice of scrupulous and, so to speak, *artistic* truthfulness continually excite my admiration. It has not unfrequently happened to me to hear a remark or statement, which I had made to one of its members, alluded to by another, in which case I have been usually astonished at the precision with which it was repeated. The repetition of the statement retained the precise shade of sense that I originally intended to convey, yet it was almost always presented in a simpler and more striking form. The essentials had been truthfully adhered to; the nonessentials were pruned off and the language was improved. The rarity of a faculty like this is easily tested by the experience of the well-known game of " Russian Scandal," and has probably been impressed on most of us when we have discovered some misrepresentation of what we did or said. Truthfulness of expression adds greatly to the charm of life; it gives a grateful sense of confidence towards those who are distinguished for it and it makes conversation more real and far more interesting. There is an exact parallel between truthfulness of expression in speech and that of delineation in drawing. In the earliest sketch it is far better to be hard in outline than inaccurate. Subsequent touching up can smooth away the hardness; but there exists no proper material to work upon when there was carelessness in the first design.

CHAPTER III.

ORIGIN OF TASTE FOR SCIENCE.

Preliminary—Extracts at length—Analysis ; Innate tastes—
Fortunate accidents—Indirect motives or opportunities—
Professional duties—Encouragement at home—Influence
and encouragement of friends—Influence and encouragement of tutors—Travel in distant parts—Unclassed
residuum—Summary—Partial failures.

WHAT were the motives that first induced the men on my list to occupy themselves with science ?

A question such as this may seem hard to answer, except in very general terms. Those who are but little versed in statistics may be daunted by reflecting on the infinite diversity of characters and antecedents ; while those who are, will be less easily discouraged. Reiterated experience will have shown them how surely, in every case with which they have dealt, the great majority of causes, or what might be better named "pre-efficients," admitted of being analysed and grouped into natural orders, leaving a minority of unclassed influences, which themselves form a class of their own, and which can be reduced indefinitely, in proportion to the minuteness with which the statistician cares to pursue his analysis. The statistics of railway accidents will serve as an example. When Captain Douglas Galton was secretary of the railway department of the Board of Trade, he succeeded in sorting their causes into the groups in which we have since been accustomed to see them printed year after year. So long as the general system of management of a railway is little changed, the same statistical ratio is maintained among them, a given proportion of accidents being due to this cause, and another to that. We may therefore estimate with some certainty the saving of life and limb, or of material of various descriptions, that will be effected when any one of these causes shall be wholly or in part removed. Similarly my aim is to group the influences which first urged the men on my list to pursue what afterwards became their favourite occupation. We shall learn the relative importance of these influences, and be enabled to estimate with greater precision than before, the value of proposed methods for making the pursuit of science more common than at present.

The returns I am about to quote are replies to the following questions :—" Can you trace the origin of your interest in science in general, and in your particular branch of it? How far do your scientific tastes appear to have been innate?"

The answers were of unequal length and minuteness. From the longer ones I have extracted what was essential, and in these and in the rest I have taken a very few editorial liberties, as already mentioned.

At this stage of the inquiry it becomes advisable to separate the replies according to the branch of science pursued by those who made them. I have not kept geography separate, because there are not many geographers on my list, and those who were, admitted of being sorted under other titles. With this exception the divisions I have adopted are much the same as those of the various Sections and Sub-sections of the British Association.

Some doubt may be felt as to how far the replies may be trusted. For my own part, I believe they are substantially correct, judging principally from internal evidence, and partly from having questioned different members of several families, and finding their opinions corroborative. The greatest difficulty I have had in my inquiries generally is due to reticence on the part of the writers, who say nothing when much was to be said; but even this does not affect *relative* results. Again, many men are conceited; still the forms in which conceit shows itself do not much affect those results. Thus, a too emphatic narration of early achievements does not distort their mutual proportions. If men are too proud to acknowledge their indebtedness to natural gifts, the relative value they ascribe to motives remains unchanged. I am astonished at the unconscious vanity which

I have elsewhere met with when making inquiries in heredity, shown by men who, owing enormously to natural gifts, wish to accredit their own free will with being the real causes of their success. One phase of this form of vanity is prominently illustrated by the late John Stuart Mill, in his strange and sad autobiography, who declares (p. 30) that he was rather below par in quickness, memory, and energy, and that any boy or girl of average capacity and healthy physical constitution, who was properly taught, could make as rapid progress in learning as he did himself! As regards the scientific men, I find, as I had expected, vanity to be at a minimum, and their returns to bear all the marks of a cool and careful self-analysis. My bias has always been in favour of men of science, believing them to be especially manly, honest, and truthful, and the results of this inquiry has confirmed that bias.

The influences and motives which urged the men on my list to occupy themselves with science fall under the heads given below. I have distinguished each head by a letter, and added to each reply the letters that seemed appropriate to its contents. The replies are subsequently analysed according to these letters.

SIGNIFICATION OF THE LETTERS.

Number of Instances.

- *a.* 59 Innate tastes (*mem :* not necessarily *hereditary*).
- *b.* 11 Fortunate accidents. It will be noticed that these generally testify to the existence of an innate taste.
- *c.* 19 Indirect opportunities and indirect motives.
- *d.* 24 Professional influences to exertion.
- *e.* 34 Encouragement at home of scientific inclinations.
- *f.* 20 Influence and encouragement of private friends and acquaintances.
- *g.* 13 Influence and encouragement of teachers.
- *h.* 8 Travel in distant regions.
- *z.* 3 Residual influences, unclassed.

EXTRACTS AT LENGTH.

PHYSICS.

(1) "My tastes are entirely innate ; they date from childhood." (*a*)

(2) "As far back as I can remember, I loved nature and desired to learn her secrets, and have

spent my whole life in searching for them. While a schoolboy I taught myself botany, chemistry, &c. . . . . under great difficulties. I had no teacher except a kind apothecary, whose knowledge was limited." (a)

(3) "From a youth, I always preferred the man of marked ability to the man of action alone. Thrown for so many years of my professional life among men chiefly of the latter class, and my sympathies being more drawn towards those in the decided minority, my tastes were, I conceive, not acquired but innate. In the early days of my professional career I gained the friendship of . . . . , of the highest professional standing, whose acquired general knowledge and love of science and observation were far beyond those of the ordinary . . . . of his time. I was both his young friend and favourite assistant for three years. He imbued me with his respect for science, and formed my character for earnestness and accuracy. . . . . To some extent, my tastes were determined by events after manhood; because in . . . extend-

ing over ten years, I held positions of great responsibility [in different parts of the world], but I consider my scientific tastes were formed in youth, that is, from 16 to 21 years of age." (*a, f, h*)

(4) "From an early age I was addicted to mechanical pursuits. In the last few years of my schooldays I took to chemistry. Entered . . . . college, expecting, after two or three years there, to [join a relative's] business as calico-printer, and gave especial attention to chemistry on that account. . . . . I had never attended *specially* to physics until appointed professor of natural philosophy. . . . . [This and subsequent similar advancement] determined me to devote myself thenceforward definitely to physics, and not to try for a chemical appointment . . . " (*a, d*)

(5) "Naturally fond of mechanics and of physical science, in which all my study has taken the direction of those departments bearing on . . . . , owing to my feeling that through the possession of special instruments for investigations in it, I could work to greater advantage;

not from any natural preference for . . . . over the other departments of physical science." (*a, c*)

(6) " My tastes were partly natural, partly encouraged by an eminent friend . . . . , who had been honoured himself by the friendship of most of the leading men of science in the early part of this century." (*a, f*)

(7) [Yes.] "I remember [incidents which proved an innate taste quoted at length] before I could write, [but] I believe the origin of my pursuit of physical science was when I attended the natural philosophy class at . . . . I was intended for business, but conceiving a distaste for it, I left it and attached myself to science." (*a, g*)

(8) "I cannot say, except that I had an innate wish for miscellaneous information. My interest in science arose from the chance circumstance of my choosing civil engineering as a profession, and having spare time, when studying at . . . , which I devoted to . . . . My scientific tastes were subsequently determined by my not

having any profession, except civil engineering, which I never followed." (c)

(9) "Ocean voyaging in beginning of life. Solitary observing for years in an observatory, placed in a country verging on a desert, but under southern skies, rich in stars unknown to the ancients and not appreciated by the moderns." (d, h)

(10) "The origin of my interest in science is mainly due to my father's knowledge of geology, navigation, and engineering. My scientific tastes were confirmed by lectures, by . . . . and . . . . and . . . . and especially by the encouragement of the latter." (e, g)

(11) "Primarily derived [both by inheritance and education] from my father." (a, e)

(12) "My first start was reading a child's story, called the 'Ghost,' where a philosophical elder brother cures his younger brother of superstition, by showing him experiments with phosphorus, electricity, &c. This set me on making an electrical machine with an apothe-

cary's phial, &c. I was then about 12 years old. My grandfather had scientific tastes to some degree. My grandmother's brother . . . . was a good amateur chemist and astronomer. He was a well-known leader of musical, and to some extent, of scientific society, at . . . ." (a)

(13) "A mathematical tendency, I think, led me first towards . . . . inquiry, to which I have been faithful ever since. Professional duties and civil engineering kept up a disposition to appreciate the material constituents of the world, and led, through surveying, in the direction of physical geography. The distinct origin of my desire to place myself among scientific students was the wonderful impression produced on me by the aspect of nature, as seen in the . . . . combined with what I may call the accident of my having been allowed to explore a part of it in an official capacity. Having thus made rather large botanical and geological collections, I came to England with them, and while employed in arranging and distributing them, picked up a certain rather irregular and un-

systematic scientific education, in the company of . . . . and others. Forced back into professional life, special scientific inquiry has not been possible; but I have had opportunities of aiding the progress of science, which I have endeavoured to make the best of." (*a*, *d*, *f*, *h*)

(14) "Largely determined by my service in north polar and equatorial expeditions." (*d*, *h*)

(15) "I am not aware of any innate taste for science. I can only remember in boyhood the influence of the Philosophical Society of . . . . and of a juvenile philosophical society in which I took interest. My interest in astronomy, especially, was very small indeed, until I was appointed [to the directorship of an observatory]." (*d*)

Mathematical Subsection.

(16) "I always regarded mathematics as the method of obtaining the *best* shapes and dimensions of things; and this meant not only the most useful and economical, but chiefly the most harmonious and the most beautiful. . . . . I was

taken to see . . . ., and so, with the help of 'Brewster's Optics' and a glazier's diamond, I worked at polarization of light, cutting crystals, tempering glass, &c. I should naturally have become an advocate by profession, with scientific proclivities, but the existence of exclusively scientific men, and in particular, of . . . . , convinced my father and myself that a profession was not necessary to a useful life." (a, e, f)

(17) "My taste for mathematics appears innate. As a boy, I delighted in sums. I trace the origin of my interest in general science to my acquaintance with . . . ., which dates from the time when I was about 15 years of age. I taught myself in mathematics and chemistry during my apprenticeship to a civil engineer and land surveyor, and subsequently studied . . . . [abroad]. My scientific tastes were largely developed through my first going [to the continent] with . . . ." (a, f)

(18) "An early taste for arithmetic, and in particular for long division sums." (a)

(19) [The following is an extract from biographical notes kindly communicated to me of the late Archibald Smith.] "Yachting would give an interest to all nautical matters, and the intimacy of his father with . . . . gave a bias towards magnetism. In a letter to one of his sisters (no date, ? about 1838), he says :—'. . . . told me he was going to write directions for ships, finding and allowing for the error caused by the local attraction of ships. So, for my own amusement and partly to help him, I wrote a set of instructions and gave them to him.' His mind was thus turned to the subject. I think it was natural to him to inquire into the reason of things. Fond of figures when a boy." (a, b, c, f)

(20) "My interest in mathematics began at . . . . [university], and was mainly due to the energy and encouragement of my tutor . . . . ; but Professor . . . . first inspired me with the sense of the magnificence of mathematics." (g)

CHEMISTRY.

(1) "Thoroughly innate. My first taste for chemistry dates from the possession of a chemical box, when I was a little boy. Whenever I had a chance of turning from other studies to natural science, I always turned. I liked play better than all other work, and chemistry better than play." (*a, b*)

(2) "Perhaps wholly innate. My first notions of chemistry were picked up from books, and I got the nickname of 'experimentalizer' at school. My taste for zoology arose through friendship with . . . . My tastes were largely determined by three years' voluntary work at chemistry, under Dr. . . . ." (*a, f*)

(3) "I was always observing and inquiring, and this disposition was never checked nor ridiculed in my childhood. My taste for chemistry dates from the lectures I attended as a boy, and to the permission to carry on little experiments at home in a room set apart for the purpose. I was encouraged in my tastes at home. Sub-

sequent determining events were my residing abroad, and my mother making a home for me there." (*a, b, e*)

(4) "They date from a very early period, and there was little to produce them in my early surroundings. As a small boy I was fond of reading books bearing on natural science. I was taught at home with my brothers, and was partially self-taught also. We had always the example of industry, and were encouraged to think for ourselves. I first studied chemistry at . . . . . College." (*a, e*)

(5) "From an early age I had an innate taste for all branches of natural science. As a boy, I made large collections of dried plants, minerals, beetles, butterflies, stuffed birds, &c. At . . . . I studied without regard to future profession for two years, and only took up chemistry as a special study on my third year's residence there." (*a, c*)

(6) "I cannot trace the origin. I began to study chemistry æt. 18, and pursued it at such

times as my duties in . . . . gave me leisure, and without any instructor. The obtaining of correct and accurate results in chemical analysis gave me great satisfaction." (c)

(7) " Scarcely innate. I ascribe the origin of my scientific interests chiefly to being sent as a pupil to an eminent man of science, Professor . . . . Subsequently I was a good deal abstracted from scientific pursuits by an early and lasting friendship with . . . . , who directed my thoughts to public work." (g)

(8) " I watched, at school, the building of a steam engine at a factory, and completely got up the whole engine. This gave my mind a start. . . . . My father gave me ' Henry's Chemistry ; ' that, and afterwards ' Turner's Chemistry,' were more interesting to me than any books of fiction. . . . . I believe at one time I read little else but ' Turner's Chemistry ' and books of poetry in whatever holiday I had. . . . I owe to my mother a child's curiosity and afterwards a man's reverence for scientific truth. I cannot tell if my scientific tastes were innate. The university,

inviting me to fill the . . . . chair, gave my work its bent. . . . ." (*d, e*)

(9) " I can trace my interest in chemistry to reading by accident a book upon it." (*b*)

(10) " I did nothing, even *quasi*-scientific, till after leaving college; nothing serious till æt. 23. My pursuit of chemistry is entirely due to circumstances occurring after manhood, and in direct opposition to family influences." (*z*)

(11) "To the opportunity afforded for study of science at . . . My tastes received no encouragement whatever from relations, my mother excepted." (*e, z*)

GEOLOGY.

(1) " Decidedly innate as regards coins and fossils. My father and an aunt collected coins and geological specimens, and I have both coins and specimens which have been in my possession since I was 9 years old. Subsequently my pursuits were influenced to some extent by the discoveries in . . . . , but at that

time I had already a considerable collection." (*a, c, e*)

(2) "A natural taste for observing and generalizing, developed by noticing the fossiliferous rocks which happened to occur in the neighbourhood of the school where I was. Afterwards the surgeon to whom I was articled, who had an observant mind, fostered my tastes." (*a, b, f*)

(3) "A natural taste. My interest in science began very early, originating in a love of experiment, at first in chemistry. . . . . The ultimate direction of my scientific tastes dates after the completion of my regular education." (*a, c*)

(4) "I believe I may say, innate, to a very considerable extent, not remembering that any definite steps were taken to inculcate science. I was indebted in a high degree to collections made by my father and mother, in . . . . , and to early familiarity with charts of those seas, and conversations on matters pertaining thereto. Afterwards, to going to Germany and finding in

the mining officers a body of men receiving a regular scientific education. Lastly, to a great extent by going for a winter to .... [in Germany], and by conversations with .... and ...." (a, e, f)

(5) "I was always fond of natural history; collected plants, insects, and birds, at [school] and fossils at [college], where .... 's lectures attracted me to geology, and subsequently, by the acquaintance of Professor ...., to the particular branch [of it which I have pursued]." (a, f, g)

(6) "As well as I can recollect, they were innate. I remember, as a boy of 6, seeing a spring in Lavender Hill; not being satisfied at the explanation, and determining to work it out for myself. I believe that I should have devoted myself to chemistry and physics, but that I was started, as a youth of 19, to travel 10 months out of the twelve on business, and so continued for 20 years. This led to my visiting all Great Britain, and to great opportunities for geologising and determined me to that study. I

worked hard at business all day (a very anxious business), and at evening and night would work hard at chemistry and geology. I found a wonderful relief in science." (*a, c*)

(7) "I believe the desire for information and habits of observation to be in a great measure innate. They were first developed by a little elementary teaching in physics and chemistry, at school, æt. 7–13. I worked alone at science at home, from the age of 11 years, where I was encouraged by the example of an elder brother. Subsequently, my pursuits were much influenced by being thrown, at an early age, æt. 19, on my own judgment and resources. I founded a mining colony in the backwoods of . . . . , and had to carry it out with several thousand people, quite alone." (*a, e, h*)

(8) " . . . . I was always apt to observe stones closely with regard to their qualities" [but the scientific taste for geology was not developed till after manhood]. (*z*)

BIOLOGY.

Zoological Subsection.

(1) [Yes.] " Inherited from my father's family, who have generally been attached to natural history [especially botany—most remarkable examples are given]. My scientific tastes were largely determined by being appointed . . . . ." (*a, d, e*)

(2) " Certainly innate. . . . . Strongly confirmed and directed by the voyage in the . . . ." (*a, h*)

(3) " Love of observation and natural history innate; [I had them] as early as I can remember. My grandfather was very fond of natural history, and a [more distant] relative has written an excellent fauna of . . . . The help of Mr. . . . . has aided me immensely, but not, I think, altered my tendency." (*a, e, f*)

(4) " Homology innate, and derived from my mother. I trace the origin of my interest in science decidedly to my mother's observations in

our childhood rambles, on the plants and animals we saw. She told me that crabs were 'sea-spiders,' and periwinkles (*Littorinæ*) 'sea-snails.' I feel sure she had never read 'De Maillet!' . . . ." (*a, e*)

(5) "I believe I inherited my general taste for scientific pursuits from my grandmother; but my choosing . . . . for special investigation resulted from a positive fascination which the very obscurity of the subject exerted upon my mind. It was perhaps a mere desire to unravel the marvellous. My scientific tastes were largely promoted by the attractive teaching of [. . . . various professors]." (*a, c, e, g*)

(6) "Thoroughly innate. I had no regular instruction, and can think of no event which especially helped to develop it. Bones and shells were attractive to me before I could consider them with any apparent profit, and books of natural history were my delight. I had a fair zoological collection by the time I was 15. My father had no scientific knowledge; nevertheless, he encouraged me in all my tastes, giving me

money freely for books and specimens, against the advice of friends; but he was indulgent generally, and not in the scientific direction only." (*a, e*)

(7) "Innate, as far as a love of nature and of the observation of natural phenomena. I trace the origin of my interest in science to the love of truth and of mental cultivation in my father, and his encouragement of this love in his children. I do not think it was largely determined by events after manhood." (*a, e*)

(8) "I should say innate. I caught at all scraps of lessons for self-improvement. My soon-developed enthusiasm must have been derived from my mother's family. As to whether they were largely developed by events occurring after manhood, I think not. All I can say is, that neither profession nor marriage nor sickness has been able to affect them." (*a, e*)

(9) "I cannot recollect the time when I was not fond of animals, and of knowing all I could learn about them. Living in the country, I had

abundant opportunities for indulging my taste, though, of course, I was not allowed to keep half the number of 'pets' I should have liked. The example of my father and elder brothers, who were all pretty firm to field sports, was also followed by me, and from field sports to field natural history is but a step. I obtained, by a piece of sheer good luck, the travelling fellowship of . . . ; it was tenable for nine years, and its income was sufficient to keep me during that time without being obliged to enter any profession. Though circumstances subsequently interfered with my using this assistance to the most advantage, in gratifying my taste for natural history, it was enormously furthered thereby." (*a, b, c, e*)

(10) "My partiality for the natural history sciences was initiated partly by my selection of medicine as a profession, and perhaps even more that, during the period of my apprenticeship, I was much under the influence of a remarkable man . . . , a most accomplished naturalist and of singularly independent judgment . . . For

three years I spent every Sunday morning with him. During this time he was constantly stimulating me (a willing follower) to work in his department of natural science, and at the same time, ever inculcating a spirit of scientific scepticism." (*d, f*)

(11) "To love of birds, their study, their dissection. I remember trying to find out in the structure of the oviduct the cause of colour and markings in the different eggs. I discovered hairs sticking in the cuckoo's stomach, arranged in a spiral manner, before I knew that John Hunter had described the same. Then I took to drawing skulls and skeletons, and my fate was sealed. That I inherited a strong love of nature is certain, from my father, who was devoted to horticulture and very fond of birds and of landscape scenery; but I cannot trace any direct tendencies or work on the part of any member of my family, except my brother. I feel that I must have had a taste for science, independently of external circumstances. At the age of 17 or 18, I had dissected every new kind of bird that I

met with. Later opportunities were entirely made by myself, or perhaps, rather, taken advantage of by myself." (*a, e*)

(12) "My love of natural history (so common in boys) showed itself in collecting insects, shells, and birds' eggs, and delighting in reading such books as Stanley on Birds, White's Selborne, Waterton, &c., at a very early age (8 years or before), and being rather encouraged than checked, continued to grow till it developed into a fondness for anatomical pursuits generally, which was never abandoned. My taste [for science] was entirely innate; no [other] member of the family nor early friend or acquaintance had any special taste for any of the natural history sciences. Two brothers, of nearly the same age, and with precisely the same surroundings, though joining occasionally in some of the above-mentioned boyish pursuits, never pursued them with real interest, and soon entirely gave them up." (*a, e*)

(13) "As a boy, I had no taste for natural history, but a passion for mechanical contri-

vances, physics, and chemistry. I earnestly desired to be an engineer, but the fact that I had a . . . . [near relative] a medical man, led to my being apprenticed to him, and I took to physiology and anatomy, as the engineering side of my profession. [The inclinations above mentioned were] altogether innate, and, so far as I know, not hereditary; neither of my parents nor any of the family showing any trace of the like tendencies. My appointment to the surveying ship . . . . made me a comparative anatomist, by affording opportunities for the investigation of the structure of the lower animals. My appointment to . . . . forced me to palæontology." (*a, c, d, h*)

(14) "My school nickname was 'Archimedes;' I was always fond of construction. If I had followed my own bent, I should probably have been [successful as] an engineer. My turn for scientific inquiry led me in early life to systematise and generalise the knowledge of others. Latterly I have felt more interest in original investigations." (*a, c*)

(15) " I was in a general atmosphere of scientific thinking and discipline. My taste for biology began with keeping insects; for chemistry and physics, by being led to try experiments. Largely inherited from my father. I have made my circumstances more than they have made me." (*a, c, e*)

(16) " My father's example influenced me so early that I have no means of judging, but I doubt much their innate character. Their origin was due primarily, beyond all probability of doubt, to my father's influence and example. They were not influenced by subsequent events, but the tastes once planted rather determined the events. My medical profession caused me to suspend my scientific pursuits for some years; but the accidental perusal of . . . . brought me back again to the study of the . . . . , and all the rest followed in due time." (*b, e*)

(17) " They appear to have been inherited. My interest in science arose from the example of my father, and the fact of my being for a year the assistant and close companion of Pro-

fessor .... of ..... at whose side I visited the poor in the lanes of .... , day and night. First began to work and concentrate energies to one branch æt. 21, when appointed ...."
(*a, d, e, g*)

(18) "They have been, I believe, nearly in an equal degree the mixed result of a natural bias and education, and were determined by professional study, when a love of scientific knowledge for its own sake first took possession of my mind." (*a, d*)

(19) "How far innate, and how far acquired and developed from my early youth, I cannot say. My love for animals of all kinds was very strong, and to gratify it I overcame every obstacle put in my way at home, when I was a boy. I trace the origin of my interest in science to the earliest impressions of my childhood, all of which, so far as I recollect them, are connected with my father, and the various animals he brought me as pets. They were not largely determined by events after manhood. I should have been an observer of animal life

under any conditions under which I might have lived." (*a, e*)

(20) "I cannot trace the origin of my interest in geology. I believe it to have been innate. I began collecting birds and studying them before I went to school, and without any inducement. I was always told by my relations that my scientific pursuits would stand in my way, but adhered to them notwithstanding. They were not at all determined by events occurring after I reached manhood; they simply increased as I grew older." (*a*)

(21) "I perceive no evidence of their being innate [? hereditary], unless I derived any tendency from my mother, who was at one time much with her great-uncle [. . . . the founder of one of our great industries] and greatly interested in his pursuits. She worked a good deal at chemistry, and was well acquainted with many of the processes in pottery. I belonged to an industrious family and saw everyone working. The attraction I have for chemistry (which is a strong one, only my profession has never

allowed me to follow it very closely) arose from being sent to work, æt. 15, in a chemical laboratory." (e)

(22) "I do not consider them innate, but induced by the following circumstances:—When I was at school, æt. 13–15, a lady, an old friend of my mother, gave me a few British shells, with their names, and a copy of 'Turton's Conchological Dictionary.' I thenceforth diligently collected British shells, and afterwards extended my researches." (b)

(23) "To my father's example (in science); to the profession of medicine (in physiology, anatomy, and . . . .). It was my interest in my profession to work at scientific subjects, while young and while waiting for practice. The example of many men whom I knew when young proved a great stimulus and incentive." (e, d, f)

(24) "Not at all innate. I can trace it distinctly to my intercourse with certain professors . . . . ; subsequently to my desire to investigate

certain scientific questions bearing on medicine, and later to my intercourse with . . . . and . . . ." (*c, d, f, g*)

BIOLOGY.

Botanical Subsection.

(1) "My scientific tastes were inborn" [and strongly hereditary]. (*a*)

(2) "As far as the word applies in any case, I should say decidedly innate. Excepting such influence as a little encouragement at home, I am unable to trace any external stimulus. At æt. 6, I was given Joyce's 'Scientific Dialogues,' which I soon mastered, then other books; before æt. 8, I commenced making star maps; æt. 12–13, I made some geological sections with tolerable correctness; and so on. It [then] seemed as if any accident and the love of new vistas were enough to lead me from one branch of science to another." (*a*)

(3) "Always fond of plants." (*a*)

(4) "Was always fond of objective and experimental knowledge. I date my first efforts of any

consequence from an early intimacy with Professor . . . , whose pupil and assistant I was. I had a fondness for science before, but the necessity for accurate and rigid observation then first dawned upon me. Subsequent events were going to . . . . [abroad], and appointments in . . . . [a foreign country, where I was much detained indoors that] compelled me to take to the microscope and study of the lower orders of plants and animals, many of which I could grow in my own room." (a, c, g)

(5) "As a youth, I followed, of my own free will, mineralogy, chemistry, anatomy, and mechanics, but chiefly chemistry. My tastes were certainly not hereditary. They were directed to botany purely through accidental circumstances [which led to a prolonged residence in an imperfectly civilized country]. I examined its plants, then wholly unknown to Europeans, but was at that time wholly ignorant of the very elements of botany. Was subsequently encouraged by . . . [eminent botanists of the day]; went to and from England and made extensive collections.

My wife actively assisted me in my botanical and other scientific pursuits, and to her advice and assistance I owe much of my success in life." (*a, f, h*)

(6) " The love for botany was instilled into me in very early youth by my father. We lived in the house of . . . . [a very eminent geologist], in the vicinity of . . . . , and I often took walks to those hills and collected plants. I also cultivated plants in our garden. A taste for natural science, especially botany, seems to have been innate. The companionship of . . . . incited me to prosecute botany with vigour. I was one of his best pupils, and travelled over a great part of . . . . with him." (*e, g*)

(7) [A posthumous account.] " He appears to have been attached to natural history all his life through, but never took up botany to any extent till the professorship was vacant. [There is some conflict of testimony here.] I think his scientific tastes were innate. I have excellent drawings of insects made by him as a schoolboy; also, he made a model of a caterpillar; tried a

little chemistry; made lace with bobbins of his own contriving. . . . . It was said, 'Nothing escapes that boy's eyes.'" (*a, d*)

(8) "To my father's encouragement of a natural inclination." (*a, e*)

(9) "I cannot trace the origin of my interest in any particular branch of science further than that as far as regards . . . . botany, I was thrown into the society of a gentleman who took much interest in it. My scientific tastes originated, as a matter of fact, after leaving . . . . [the university]." (*f*)

(10) "Not innate. I trace the origin of my botanical tastes to leisure; to the accidental receipt of De Candolle's 'Flore française,' whilst resident in that country; and to encouragement from my mother. They were determined afterwards by independence (considering my absence of ambition to rise in the world) and by friendship and encouragement from . . . . , the four greatest British botanists of the day." (*b, e, f*)

BIOLOGY.

Medical Subsection.

(1) "Innate in a great degree. I trace the origin of my interest in science (1) to my mother's mental activity and love of collecting and arranging, and my father's constant encouragement of my pursuits; (2) to the friendship of [three eminent botanists], by whom I was chiefly induced to study botany; (3) to my profession, the choice of which was in some measure determined by my taste for collecting and studying." (*a, d, e, f*)

(2) "I selected the medical profession because it was that of my father. This choice led me to scientific pursuits, for which I had no previous predilection, as I had no opportunities that way. I conclude the tastes were innate, as they certainly showed themselves the moment the opportunity for developing them occurred, namely, at the commencement of my professional studies, æt. 17." (*a, d*)

(3) "Not at all especially innate. I could have taken to any other subject quite as well, so far as I know. I trace the origin of my interest in science to the knowledge that I must do my best in it to earn a livelihood and to please my parents. I did not follow my own branch from any special liking—indeed, I disliked it; but it was necessary to follow some branch. The connection with an hospital and medical school in . . . . have been inducements to continue work, and all my life I have worked pretty steadily." (d)

(4) "I cannot perceive that they were innate. Possibly my tastes were due to retentiveness of memory as to objects and facts, and a strong impression that good surgery is a great fact. Subsequently, by the approval of teachers, when between æt. 18 and 20, having been selected chief assistant to the most popular teacher of anatomy of his day, and also to a professor of surgery." (c, g)

(5) "Had an interest excited in philosophical

inquiries by my father's acute observations in all such topics." (*e*)

(6) "I cannot say that I had naturally a turn for any pursuit in particular. My addiction to medicine was purely the result of accident. I never gave a thought to physic as a subject of study until I was 27 years old." (*d*)

(7) "Accidentally [directed] to medicine by associating with a medical friend in a superficial study of botany." (*c, d*)

STATISTICS.

(1) "Certainly my scientific tastes appear to me to have been, so to say, innate." (*a*)

(2) "My interest in science was due to my having been officially employed in the early part of [my career, in a very important statistical inquiry]." (*d*)

(3) "Innate, I think. I inherit many mental peculiarities and talents from my paternal grandfather, amongst which is a love of figures and

tabulation; none from my father. I cannot [otherwise] trace the origin of my interest in science, nor were my tastes largely determined by events after manhood." (*a*)

(4) "I should be much inclined to think there was an innate tendency, but that the tastes were developed by a good and for the most part suitable education. When at my first school, æt. $10\frac{1}{2}$–12, the head-master gave very clear occasional lessons in moral and economical subjects. I can remember vividly to the present day the impression which those lessons made upon me. As I am not aware that the other boys in the class were equally impressed, I think I must have had an innate interest in those subjects; but the lessons probably increased the interest very much." (*a, b, g*)

(5) "I cannot distinguish between what I may have derived from nature and what I may have acquired from intercourse with my father and certain of his friends. When I was 11 years old, my father gave a series of lectures on electricity, mechanics, astronomy, and pneuma-

tics, to all of which, but especially to the last, I paid delighted attention. I presently began to construct apparatus for myself. Subsequently practice in teaching led me to seek for knowledge. Intercourse with men of higher attainments became a great spur; my turn for . . . . was favoured by my opportunities as an early member of the . . . . Society." (*a, e, f*)

(6) " Professor . . . . 's lectures on geology were the origin of my interest in that science; the work of the . . . . statistical society in educational inquiries influenced my taste for statistical science; frequent attendance at meetings of the British Association encouraged my scientific tastes.". (*d, g*)

MECHANICAL SCIENCE.

(1) " If any tastes be innate, mine were; they date from beyond my recollection. They were not determined by events after manhood, but, I think the reverse; they were discouraged in every way." (*a*)

(2) "Decidedly innate. The science of .... was well taught at the university of ...., where I studied, æt. 16-18, and accidentally this became serviceable to me when employed as an engineer by .... The friendship of .... materially affected my career. My tastes were not largely developed by events occurring after manhood." (*a, b, d, f*)

(3) "Family tradition derived through my mother's side. My profession fell in with my natural tastes, such as sketching." (*c, d, e*)

(4) "Innate, I think, as regards certain qualities of mind, which led me, under the pressure of circumstances, to direct my attention to certain things in a certain way, namely, (1) independence of judgment; (2) earnestness of purpose; (3) a practical, clear-headed, common sense, logical way of viewing things." (*c, d*)

(5) "I cannot say whether they were innate. I was always brought up in a half-scientific, half-literary atmosphere, and was a fair mathematician as a boy, as well as a fair classic and

linguist. My tastes were not determined by after events, but my avocations were rather determined by my scientific habits." (*e*)

ANALYSIS OF REPLIES.

Having given the replies in gross, it now becomes our business to sort their contents under different heads. It would be useless and even embarrassing to make lengthy extracts from them; short abstracts will therefore be given, which the reader may verify whenever he pleases by the help of the reference number, printed in parentheses ( ), which is the same both here and in the original.

§ A. INNATE TASTES.

Instances of a strong taste for science being decidedly innate. I have not included among these the whole of the cases to which an *a* has been affixed :—

Physics and Mathematics.—12 cases out of 20 replies. (1) My tastes are entirely innate;

they date from childhood. (2) As far back as I can remember, I loved Nature and desired to learn her secrets. (3) Always attracted by men of ability. (4) From an early age I was addicted to mechanical pursuits; then to chemistry. (5) Naturally fond of mechanics and physical science. (6) My tastes were partly natural, partly encouraged. (7) I remember [incidents which proved an innate taste] before I could write. (8) I had an innate wish for miscellaneous information. (11) Primarily derived [both by inheritance and education] from my father. (16) I always regarded mathematics as the method of obtaining both the most useful and the most harmonious, &c. (17) My taste for mathematics appears innate; as a boy I delighted in sums. (18) An early taste for arithmetic, and in particular for long division sums.

Chemistry.—5 cases out of 11. (1) Thoroughly innate. (2) Perhaps wholly innate. (3) I was always observing and inquiring. (4) They date from a very early period, and there was little to produce them in my early surroundings. (5) From an early age I had an innate taste for all branches of science.

Geology.—At least 7 out of 8 cases. (1) Decidedly innate. (2) A natural taste for observing and generalizing, developed. (3) A natural taste ; my interest in science began very early. (4) I believe I may say innate to a very considerable extent. (5) I was always fond of natural history. (6) As well as I can recollect, they were innate. (7) I believe the desire for information and habits of observation to be in great measure innate.

Zoology.—18 cases out of 24. (1) [Yes.] Inherited from my father's family. (2) Certainly innate. (3) Love of observation and natural history innate. (4) Homology innate. (5) I believe I inherited my general taste for scientific pursuits. (6) Thoroughly innate. bones and shells were attractive to me before I could consider them with any apparent profit. (7) Innate love of nature and observation of natural phenomena. (8) I should say innate ;

I caught at all scraps of lessons for self-improvement. (9) I cannot recollect the time when I was not fond of animals and of knowing all I could learn about them. (11) Love of birds and their study . . . I feel that I must have had a taste for science independently of external circumstances. (12) My taste [for science] was entirely innate. (13) As a boy I had a passion for mechanical contrivances; [my scientific tastes are] altogether innate. (14) I was always fond of construction; my turn for scientific inquiry led me in early life to systematise the knowledge of others. (15) Largely inherited from my father. (17) They appear to have been inherited. (18) Nearly in an equal degree the mixed result of a natural bias and education. (19) I should have been an observer of animal life under whatever conditions I might have lived. (20) I believe my interest in zoology to have been innate.

Botany.—8 cases out of 10. (1) My scientific tastes were inborn. (2) As far as the word applies in any case, I should say decidedly innate. (3) Always fond of plants. (4) Was always fond of objective and experimental knowledge. (5) As a youth I followed of my own free will chemistry and other sciences. (6) A taste for natural science, especially botany, seems to have been innate. (7) [Scientific tastes apparently innate.] (8) A natural inclination.

Medical Science.—Only 2 cases out of 7. (1) Innate in a great degree. (2) I conclude the tastes were innate, as they showed themselves the moment the opportunity for developing them occurred.

Statistics.—3 cases out of 6. (1) Certainly my scientific tastes appear to me to have been, so to say, innate. (3) Innate, I think. (4) Much inclined to think there was an innate tendency.

Mechanical Science.—At least 2 cases out of 5. (1) If any tastes be innate, mine were; they date from beyond my recollection. (2) Decidedly innate.

INSTANCES OF TASTES BEING DECIDEDLY NOT INNATE.

Physics and Mathematics.—1 case out of 20. (15) I am not aware of any innate taste for science.

Chemistry.—1 case out of 11. (10) I did nothing serious till æt. 23. My pursuit of chemistry is entirely due to circumstances occurring after manhood.

Zoology.—3 cases out of 24. (16) I doubt much their innate character. (22) I do not consider them innate, but induced. (24) Not at all innate.

Botany.—1 case out of 10. (10) Not innate.

Medical.—4 cases out of 7. (3) Not at all especially innate. (4) I cannot perceive that they were innate. (6) I cannot say that I had naturally a turn for any pursuit in particular. (7) Accidentally [directed] to medicine.

Statistics.—1 at most out of 6. (2) My interest in science was due to my having been officially employed in a statistical inquiry." [It is with much hesitation that I consent to enter this as a case of " not innate."]

SUMMARY OF RESULTS AS TO INNATE TASTES.

	Total cases.	Decidedly innate.	Decidedly not innate.	Doubtful.
Physics and Mathematics	20	12	1	7
Chemistry and Mineralogy	11	5	1	5
Geology	8	7	0	1
Biology—Zoology	24	17	3	4
Botany	10	8	1	1
Medical Science	7	2	4	1
Geography (not discussed separately)	0	0	0	0
Statistical Science	6	3	1	2
Mechanical Science	5	2	0	3
	91	56	11	24

A mere glance at the table and at the foregoing extracts will probably be enough to convince the reader that a strong and innate taste for science is a prevailing characteristic among scientific men; also that the taste is enduring. This latter peculiarity is by no means a necessary consequence of the former;

on the contrary, the ruling motives in the disposition of a man usually change as he grows older, the love of inquiry in childhood being superseded by the fierce passions of youth, and these by the ambitions of more mature life. But a special taste for science seems frequently to be so ingrained in the constitution of scientific men, that it asserts itself throughout their whole existence. Obviously it must have had great influence in directing their early studies and in ensuring their successful prosecution of them in after years.

It would be a curious inquiry to seek the limits of a special taste, that is the diversity of the objects, any one of which would satisfy it. I think the indications are clear that the tastes of some of my correspondents are far more special than those of others, and that the latter have checked a tendency to desultoriness by their strength of will, or have had it checked by the necessities of their position as professors or professional men; or, most of all, by the possession of that strange quality which the phrenologists call adhesiveness, but which seems to defy analysis. It exists in very different strength in different persons, and I know not where to find a better illustration of its power than in the ordinary case of a man falling in love for the first time. Few lookers-on will doubt that almost any young man is capable of falling in love with any one of at least one-third of the presentable young women of his race and social position, if they happen to see much of one another under favourable circumstances and without other distraction; yet, although the innate taste is of so general a character, it becomes specialised at once by the mere act of falling in love. Then the image of one woman takes complete possession of his thoughts; she is for a considerable period the only female who has attractions for him, although he might previously have been equally attracted by any one of tens of thousands of her sex.

A strong taste bearing remotely on science may prove very helpful. The love of collecting, which is a trifling tendency in itself, common to children, idiots, and magpies, often leads to the study of the things collected, and is of immense use to a man who wishes to study objects that must be collected in large numbers. I have been told of an astronomer whose primary taste was a love of polished brass instruments and smooth mechanical movements, that nothing satisfied this taste so fully as work with telescopes, and from loving the instruments he soon learnt to love the work for which they were used. A taste for careful drawing works well into engineering and into systematic botany or zoology. A love of adventure and field sports may be an extremely useful element in the character of a man who follows geology or zoology.

As a rough numerical estimate, it seems that 6 out of every 10 men of science were gifted by nature with a strong taste for it; certainly not 1 person in 10, taken at haphazard, possesses such an instinct; therefore I contend that its presence adds five-fold at least, to the chance of scientific success. The converse way of looking at the question gives a similarly large estimate. Certainly one-half of the population have no care for science, and an extremely small proportion of that half succeed in it. Nay, further, it appears (though I cannot publish facts in evidence, without violating my rule of avoiding personal allusions) that of the men who have no natural taste for science and yet succeed in it, many belong to gifted families, and may therefore be accredited with sufficient general abilities to leave their mark on whatever subject it becomes their business to undertake. We may therefore rest assured that the possession of a strong special taste is a precious capital, and that it is a wicked waste of national power to thwart it ruthlessly by a false system of education. But I can give no test which shall distinguish in boyhood between a taste that is destined to endure and a passing fancy, further than by remarking that whenever the aptitudes seem hereditary, they deserve peculiar consideration.

Instinctive tastes for science are, generally speaking, not so strongly hereditary as the more elementary qualities of the body and mind. I have tabulated the replies, and find the proportion to be 1 case of inheritance to 4 that are not inherited from either parent. There is no case in which the correspondent speaks of having inherited a love of science from his mother, though, of course, she may, and probably has, often transmitted it from a grand-parent. I have a curious case among the returns sent to me of a passion for heraldry characterising a great-nephew and a great-uncle, the latter of whom had died before the former was born. I have another of an eminent statistician, in whom a love of figures and tabulation was highly characteristic of his grand-parent and is very strongly marked in himself, but was wholly absent in his parent and all other known members of his small family. There have been numerous and most curious cases of a love of figures and tabulation in my own family, which richly deserve a full description. It was carried to so strange an extravagance by one of its members, a lady now deceased, that I can do no sufficient justice to her peculiarities by speaking in general terms; I ought to give pages of anecdote.

§ B. FORTUNATE ACCIDENTS.

We next come to a group of cases which imply a latent taste for science, namely, where a lifelong pursuit of it was first determined by some small accident. The previous indifference or equilibrium of the mind was unstable, a push was accidentally given, its position was wholly changed, and it rested in one of stable equilibrium. These cases are not numerous—only 10 altogether—but I put them in the second place on account of their affinity to those in the first.

Physics and Mathematics.—(19). [Refer to this.]

Chemistry.—(1) Possession of a chemical box when I was a little boy. (3) From lectures I attended when a boy. (9) To reading by accident a book on chemistry.

Geology.—(2) Fossiliferous rocks near the school where I was.

Zoology.—(9) A travelling fellowship. (16) Accidentally reading a book brought me back to scientific studies, previously suspended owing to my profession. (22) Gift, when a boy, of a box of British shells with a book to explain them.

Botany.—(10) Accidental receipt of De Candolle's "Flore française," when residing in France.

Medical Science.—None.

Statistics.—(4) Very clear occasional lectures when a boy.

Mechanics.—(2) A particular study at a university, which accidentally became of professional importance.

§ C. INDIRECT MOTIVES OR OPPORTUNITIES.

This group has also considerable affinity to group (A) and has been alluded to in the remarks appended to the extracts referring to it. It includes those cases in which the mind was partly, but not largely, deflected from its natural bent; that portion of the innate tendency which admitted of being " resolved in the direction" of the scientific pursuit, being satisfied, the remainder being wasted. These cases are not numerous—only 16 altogether—but I give them the third place for the same reason that I gave group (B) the second.

Physics and Mathematics.—(5) Possession of special instruments. (8) Choosing engineering as a profession, but not following it. (19) Love of yachting (leading to researches on magnetism of ships).

Chemistry.—(6) The obtaining of correct and accurate results in chemical analysis gave me great satisfaction.

Geology.—(1) Interest in discoveries made in . . . . . (3) A very early love of experiment and chemistry. (6) Should have followed chemistry and physics, but circumstances . . . . gave opportunities for geology.

Zoology.—(5) My choosing . . . . for special investigation was due to a positive fascination from the obscurity of the subject. (9) My father's and brother's pursuit of field sports, and thence indirectly to natural history. (13) An early passion for mechanism, which led me to take to physiology and anatomy, as the engineering side of my profession. (15) My taste for biology began with keeping insects. (24) .... subsequently to the desire to investigate certain questions bearing on medicine.

Botany.—None.

Medical Science.—(3) Connection of hospital and medical school with the place of his residence. (4) Love of facts and the impression that good surgery is a great fact.

Statistics.—None.

Mechanics.—(3) Profession fell in with natural tastes, such as sketching. (4) Innate faculties, serviceable to profession under the pressure of circumstances.

§ D.   PROFESSIONAL DUTIES.

The fourth group comprises instances in which professional duty was a principal cause of the interest first felt in scientific pursuits, or else of the energies being concentrated upon some branch of science towards which no special inclination had previously been exhibited. Two or three, of the 21 cases which I shall quote, may perhaps be thought doubtful examples and more appropriate to the preceding group ; but after all possible deductions have been made, there will remain ample evidence of the magnitude of the influence we are considering. A wise administrator, desirous, even at some cost, of promoting original investigation, would establish many professional offices of a scientific character, having responsible duties of a prominent kind attached to them. They would create much new interest in science, and would compel those who held them, to work steadily and to a purpose in scientific harness.

Physics and Mathematics.—(4) Had never

attended *specially* to physics till appointed professor of natural philosophy. This induced me to give up chemistry, and to devote myself definitively to physics. (9) Solitary observing for years [as director of an observatory]. (13) Professional duties and civil engineering . . . . ; official exploration of . . . . (14) Largely determined by service in north polar and equatorial expeditions. (15) My interest in astronomy was very small indeed, until I was appointed [to the directorship of an observatory].

Chemistry.—(8) The university inviting me to fill the chair of . . . , gave my work its bent.

Geology.—None.

Zoology.—(1) Largely determined by being appointed . . . . (10) Partly by my selection of medicine as a profession (13) My appointment to a surveying ship made me a comparative anatomist . . . . , that to . . . . forced me to palæontology. (17) First began to concentrate energies to one branch, when appointed . . . . (18) [My scientific tastes] were determined by professional study. (23) To the profession of medicine [in physiology, anatomy and . . . .] (24) Subsequently to my desire to investigate certain subjects bearing on [my profession of] medicine.

Botany.—(7) Never took up botany to any extent till the professorship was vacant. [There is some conflict of testimony here.]

Medical Science.—(1) Partly to my profession. (2) I selected the medical profession because it was that of my father; this choice led me to scientific pursuits. (3) I did not follow my own branch from any special liking—indeed, I rather disliked it, but it was necessary to earn a livelihood and to follow some branch. (6) My addiction to medicine was purely the result of accident: I never gave a thought to physic as a subject of study, until I was 27 years old. (7) Accidental to medicine.

Statistics.—(2) Due to official employment when young, in a very important statistical inquiry.

Mechanics.—(2) The science of . . . . . , which I had learnt accidentally, became serviceable to me when employed as an engineer. (3) My profession fell in with my natural tastes. (4) Pressure of circumstances.

§ E. ENCOURAGEMENT AT HOME.

Nearly one-third of the scientific men have expressed themselves indebted to encouragement at home. They received it in various ways; sometimes the influence of the parent was strong and direct, as "their origin was due beyond all doubt to my father's influence;" sometimes it was strong but general, as "I was in a general atmosphere of scientific thinking and discussion;" sometimes it went no further than indulgence, as "permission to carry on little experiments at home in a room set apart for the purpose." Under each and all of these shapes it was truly welcome, and its effectiveness may be in some measure estimated by the vastly smaller number of cases in which success was obtained in direct opposition to family influences.

Scientific studies in boyhood are apt to meet with scant favour at home; they deal too much in abstractions on the one hand, and sensible messes and mischief to furniture and clothes on the other. They lead to no clearly lucrative purpose, and occupy time which might be apparently better bestowed. These hindrances were far more seriously felt when the men on my list were young, when apparatus was hardly to be procured, and when scientific work was exceptional. I ascribe many of the cases of encouragement to the existence of an hereditary link; that is to say, the son had inherited scientific tastes, and was encouraged by the parent from whom he had inherited them, and who naturally sympathized with him.

Attention should be given to the relatively small encouragement received from the mother. I have sorted the extracts so as to permit the comparison to be easily made. The female mind has special excellencies of a high order, and the value of its influence in various ways is one that I can never consent to underrate; but that influence is towards enthusiasm and love (as distinguished from philanthropy), not towards calm judgment, nor, inclusively, towards science. In many respects the character of scientific men is strongly anti-feminine; their mind is directed to facts and abstract theories, and not to persons or human interests. The man of science is deficient in the purely emotional element, and in the desire to influence the beliefs of others. Thus I find that 2 out of every 10 do not care for politics at all; they are devoid of partisanship. They school a naturally equable and independent mind to a still more complete subordination to their judgment. In many respects they have little sympathy with female ways of thought. It is a curious proof of this, that in the very numerous answers which have reference to parental influence, that of the father is quoted three times as often as that of the mother. It would not have been the case, judging from inquiries I elsewhere made, if I had been discussing the antecedents of literary men, commanders, or statesmen, or, still more, of divines.

Physics and Mathematics.—(10) The origin of my interest in . . . . is mainly due to my father's knowledge of geology, navigation, and engineering. (11) Primarily derived [both by education and inheritance] from my father.

Chemistry.—(3) Permission to carry on little experiments at home, in a room set apart for the purpose. . . . . Subsequently residing abroad and my mother making a home for me there. (4) I was taught at home with my brothers; we had always the example of industry, and were encouraged to think for ourselves. (8) My father gave me [some books on chemistry, and] I owe to my mother a child's curiosity and afterward a man's reverence for scientific truth. (11) My tastes received no encouragement whatever from relations, my mother excepted.

Geology.—(1) My father and an aunt collected specimens. (4) I was indebted in a high degree to collections made by my father and mother. (7) I was encouraged by the example of an elder brother.

Zoology.—(9) (The example of my father and elder brothers, who were all pretty firm to field sports, was also followed by me, and from field sports to field natural history is but a step). (15) Largely inherited from my father. I was in a general atmosphere of scientific thinking and discussion. (21) I may have derived [? inherited] the tendency from my mother; I belonged to an industrious family, and saw every one working. (1) [Traditionally derived, and] inherited from my father's family [*i.e.* from father, grandfather, &c.] (6) My father had no scientific knowledge, nevertheless he encouraged me. (7) I trace it to the love of truth and of mental cultivation in my father, and to his encouragement of this love in his children. (11) That I inherited a strong love of nature from my father is certain, who was devoted to horticulture and very fond of birds. (16) Their origin was due, beyond all doubt, to my father's influence. (17) My interest in science arose from the example of my father, and . . . . &c. (19) I trace it to the earliest impressions of my childhood, all of which are connected with my father and the animals he brought me as pets. (23) To my father's example (in science). (4) Decidedly to my mother's observations in our childhood rambles. (8) My soon-developed enthusiasm must have been derived from my mother's family.

Botany.—(2) A little encouragement at home. (6) The love of botany was instilled into me in very early youth by my father. (8) To my father's encouragement of a natural inclination. (10) And to encouragement from my mother.

Medical Science.—(1) [Partly] to my mother's mental activity and love of collecting and arranging, and to my father's constant encouragement of my pursuit.

Statistics.—(5) [Partly] acquired from intercourse with my father and . . . .

Mechanics.—(5) I was always brought up

in a half scientific, half literary atmosphere. (3) Family tradition derived through my mother's side.

Two cases are mentioned in which the origin of the scientific tastes was partly due to the active assistance of the wife. One of these is Botany (5), and the other I have ventured to suppress, as it did not appear to me sufficiently decided.

§ F. THE INFLUENCE AND ENCOURAGEMENT OF FRIENDS.

This group has much in common with that of the indirect influences already classed under group C; it includes cases where a fortuitous acquaintance has been the means of deciding a career, probably by revealing a latent taste, or showing how some obstacle in the way of indulging it could easily be removed. There is a wide interval, often very difficult to get over, between the study of a subject out of books and the practical investigation of it for oneself. At this point of a man's mental progress the help of a friend may be of immense assistance; he may give elementary hints which will remove formidable difficulties to a beginner, who is utterly unused to experiment. It is told, I think, of a scholar, that he laboured for successive days to make with his own hands in his own chambers a plum-pudding according to a time-honoured family recipe, but he produced nothing except thick pastes or stirabouts of different degrees of lumpiness, revolting to the sight. At length he confided his difficulties to a lady, who explained that in making plum-puddings it was a matter of course, and therefore not spoken of in the recipe, to put the ingredients into a bag before beginning to boil them. The example of a friend encourages a young man to overcome his diffidence and to firmly occupy any position that he knows by his own judgment to be true. Perhaps the greatest help of all is the consciousness of strength which is given by co-operation on not very unequal terms with a veteran in performance and reputation.

Out of the 91 cases, 18 speak gratefully of the influence and encouragement of friends.

Physics and Mathematics.—(3) . . . . I was both his young friend and assistant for 3 years. He imbued me with his respect for science, ... , earnestness, and accuracy. (6) Partly encouraged by an eminent friend. (13) Picked up an unsystematic education [in science] in the company of . . . . (16) I was taken to see . . . . [which was the origin of my experimentalising]. (17) I trace it to my acquaintance with . . . . and to going abroad with him. (19) The intimacy of his father with . . . . gave a bias towards magnetism.

Chemistry.—(2) My taste for zoology arose through friendship with . . . .

Geology.—(2) The surgeon to whom I was articled fostered my tastes. (4) To mining officers in Germany; to conversations with . . . . and . . . . , and acquaintance of . . . . (5) Through the acquaintance of . . . . , to the particular branch [of geology, that I have pursued].

Zoology.—(3) The help of . . . . has aided me immensely. (10) I was much under the influence of a remarkable man, a most accomplished naturalist. (23) The example of many men whom I knew when I was young, proved a great stimulus and incentive. (24) I can trace it distinctly to my intercourse with certain professors.

Botany.—(5) . . . . was subsequently encouraged by [eminent botanists]. (9) I was thrown into the society of a gentleman who took much interest in botany. (10) They were determined afterwards by . . . . and the friendship and encouragement of the four greatest British botanists of the day.

Medical Science.—(1) [Partly] to the friendship of three eminent botanists. (7) Accidentally [directed] to medicine by associating with a medical friend in a superficial study of botany.

Statistics.—(5) [Partly] from intercourse with my father and certain of his friends.

Mechanical Science.—(2) The friendship of . . . . materially influenced my career.

§ G. INFLUENCE AND ENCOURAGEMENT OF TUTORS.

This group of 13 cases refers to the influence and encouragement of masters, tutors and professors. It is a small one; not because persons in those positions are incapable of exerting much salutary influence, but because the scientific men on my list seldom had the advantage of receiving congenial instruction. This is clearly proved by a comparison of the replies referring to Scotch and to English tuition. In Scotland the university programme and the general method of teaching is much more suited to men of a scientific bent of mind than those in England; consequently the influence of tutors has been testified to far more abundantly by those men on my list who have been educated in Scotland than by the rest. The proportions are striking and instructive. I find that about one-sixth of those from whom I have received returns have studied in Scotland; hence, if professorial influences had been equally efficacious on both sides of the Tweed, there would have been 5 times as many expressions of gratitude to English teachers as to Scotch. But the facts show that no less than 8 out of the 13 cases refer to teachers in Scotland, 1 to a Scotch teacher settled in England, and only 4 to English professors. It would have been (8 × 5 =) 40 and not 4, if the English education had been as profitable to science as the Scotch. I willingly admit that the smallness of the numbers, namely, only 13 cases, renders precise figures open to question; however, the superiority of the Scotch system is supported by other evidence which I shall speak of in the chapter on education.

Physics and Mathematics.—(7) I believe the origin was when I attended the natural philosophy classes at . . . . (10) Tastes confirmed by lectures, and especially by the encouragement of [certain professors]. (20) Interest in mathematics due to the encouragement of . . . . , and influence of [professors at a university].

Chemistry.—(7) Chiefly to being sent as a pupil to an eminent man of science.

Geology.—(5) Lectures by . . . .

Zoology.—(5) My scientific tastes were largely promoted by the attractive teaching of . . . . [various professors]. (17) And to being the assistant and close companion of . . . . (24) I can trace it [in part] distinctly to my intercourse with certain professors.

Botany.—(4) I date my first efforts of any consequence from an early intimacy with . . . . . , whose pupil and assistant I was; the necessity of accurate work then dawned upon me. (6) The companionship of . . . . incited me to prosecute botany with vigour; I was one of his best pupils, and travelled with him.

Medical Science.—(4) Subsequently by the approval of teachers, having been selected chief-assistant.

Statistics.—(4) Very clear occasional lectures, when a boy, on moral and economical subjects;

the tastes were afterwards developed by a good education. (6) Professor . . . .'s lectures were the origin of my interest in geology [It was the earliest scientific pursuit of this correspondent].

Mechanical Science.—None.

§ H. TRAVEL IN DISTANT PARTS.

There are only 8 cases in this group, namely, those in which the aspects of nature under new conditions have developed a love for science. Few men of scientific training have had opportunities of distant travel, but on those few their action has been very strong, especially as regards biologists and physicists. I say nothing here in respect to mere geographers, and quote none of their replies, because its importance to them requires neither proof nor comment. Men are too apt to accept as an axiomatic law, not capable of further explanation, whatever they see recurring day after day without fail. So the dog in the back yard looks on the daily arrival of the postman, butcher, and baker as so many elementary phenomena, not to be barked at or wondered about. Travel in distant countries, by unsettling these quasi-axiomatic ideas, restores to the educated man the freshness of childhood in observing new things and in seeking reasons for all he sees.

I believe that a handsome endowment of travelling fellowships, thoroughly well paid, with extra allowance for any special work allotted to their holders, given only to young men of high qualifications, and lasting for at least 5 years, would be money well bestowed in the furtherance of science.

Physics and Mathematics.—(3) To some extent my tastes were determined by events after manhood, because for 10 years I held positions of great responsibility [in distant parts of the world], but I consider they were formed in my youth. (9) Ocean voyaging in the beginning of life; solitary observing for years in a country verging on a desert, under southern skies. (13) The distinct origin . . . . was the wonderful effect produced on me by the aspects of nature, as seen in the . . . . , combined with what I may call the accident of having been allowed to explore part of it in an official capacity. (14) Largely determined by my service in north polar and equatorial expeditions.

Chemistry.—None.

Geology.—(7) Subsequently much influenced by being thrown, at æt. 19, on my own judgment and resources in founding a mining colony in the backwoods of . . . . and carrying it out quite alone.

Zoology.—(2) Strongly confirmed and directed by the voyage in the . . . . (13) My appointment to the surveying ship . . . . made me a comparative anatomist, by affording opportunities for the investigation of the structure of the lower animals.

Botany.—(5) They were directed to botany purely through accidental circumstances [which led to a prolonged residence in an imperfectly civilized country].

§ Z. UNCLASSED RESIDUUM.

We now come to the final group, namely, those influences which cannot be sorted into any of the 8 groups with definite titles, which we have already examined. At the outset I spoke of these unclassed conditions as forming a class by themselves, of no great importance, and which might be indefinitely reduced in proportion as we chose to pursue our analysis. I estimate that the 91 replies which I have received and analysed assign a total of 191 causes. It now appears that no less than 188 of these fall into one or other of 8 definite groups, and that there remain only 3 on our hands for the unclassed residuum. Even these are apparently due to aggregates of conditions, the more important of which would probably find their place among the 8 groups, leaving a still minuter residue. We may lightly dismiss them as of inappreciably small importance in our present inquiry.

Chemistry. — (10) Entirely due to circumstances after manhood, and in direct opposition to family influences. (11) To opportunity at [a foreign university].

Geology.—(8) The tastes developed gradually after manhood.

SUMMARY.

If we take a general survey of our national stock of capabilities and their produce, we see that the larger part is directed to gain daily bread and necessary luxuries, and to keep the great social machine in steady work. The surplus is considerable, and may be disposed of in various ways. Let us now put ourselves in the position of advocates of science solely, and consider from that point of view how the surplus capabilities of the nation might be diverted to its furtherance. How can the tastes of men be most powerfully acted upon, to affect them towards science?

The large category (A) of innate tastes is practically beyond our immediate influence; but though we cannot increase the national store, we need not waste it, as we do now. Every instance in which a man having an aptitude to succeed in science, is tempted by circumstances which might be controlled, to occupy himself with subjects of less national value, is a public calamity. Aptitudes and tastes for occupations which enrich the thoughts and productive powers of man are as much articles of national wealth as coal and iron, and their waste is as reprehensible. Educational monopolies which offer numerous and great prizes for work of other descriptions, have caused enormous waste of scientific ability, by inducing those who might have succeeded in science, to spend their energies with small effect on uncongenial occupations. When a pursuit is instinctive and the will is untaxed, an immense amount of work may be accomplished with ease. Witness, to take an extreme case, the sustained action of the wholly involuntary muscles. The heart does its work unceasingly, from birth to death; and it is no light work, but such as the arm, working a pump-handle, would soon weary of maintaining;

or again, think of the migratory flight of birds, in obedience to an instinct; or of the muscular force, astonishing both in magnitude and endurance, exhibited by lunatics, who have some real though morbid passion which goads them to exercise it. We must therefore learn to respect innate tastes, which directly, as in A, or indirectly, as in C, serve the cause of science. As regards B, the fortunate accidents, we can multiply opportunities. There is great hope in respect to D, the professional influences. It is clear to all who have knowledge of the scope of modern science, that there exists an immense deal of national work which has to be performed, and which none but men of scientific culture are qualified to undertake. Scientific superintendence is required for all kinds of technical education, for statistical investigations of innumerable kinds, and deductions from them; for sanitary administration in the broadest sense; for agriculture, mining, industrial occupations, war, engineering. There is everywhere a demand for scientific assessors, who shall discover how to economise effort and find out new processes and fruitful principles. Professional duties generally, ought to be more closely bound up with strictly scientific work than they are at present; and this requirement would tend to foster scientific tastes in minds which had little inborn tendency that way. In respect to G, the influence and encouragement of tutors, seeing how far Scotland has surpassed England in the attractiveness of her mode of teaching, which is by professorial lectures rather than by class-work, it is clear that the English system admits of being greatly improved, and the influence of her teachers proportionately increased, in turning the minds of youths to science. Lastly, as regards H, travel in distant lands, its indirect value deserves far more than the moderate sums assigned to its prosecution, in the way of starved travelling fellowships and rare voyages of surveying ships.

To sum up in a few words: it seems to me that the interpretation to be put on the replies we have now been considering, is that a love of science might be largely extended by fostering, and not thwarting, innate tendencies, by the extension of scientific professional appointments and professorships, by assimilating in some cases the English system of teaching to that of the Scotch, and by creating travelling and other fellowships which shall enable their holders to view nature in various aspects, and to work with foreigners whose habits of thought are fruitful in themselves, but of a different kind to our own.

I will take this opportunity of drawing attention to what appears to me one of the greatest of desiderata of this kind in the present day, namely, the establishment of medical fellowships amply sufficient to enable the best youths, who intend to follow medicine as a profession, to spend their early manhood in prosecuting independent medical researches. I appeal to capitalists, who know not what use, free from abuse, to make of their surplus wealth, to consider this want. They might greatly improve the practical skill of the English medical profession by affording opportunities of prolonged study. They might perhaps themselves, reap some part of the benefit of it. A young medical man has now to waste the most vigorous years of his life in miserable routine work simply to obtain bread, until he has been able to establish his reputation. He has no breathing-time allowed him; the cares of mature life press too closely upon his student days to give him the opportunities of prolonged study that are necessary to accomplish him for his future profession.

The influences we have been considering, are those which urge men to pursue science rather than literature, politics, or other careers; but we must not forget that there are deep and obscure movements of national life, which may quicken or depress the effective ability of the nation as a whole. I have not considered the reasons why one period is more productive of great men than another, my inquiry being limited, for the reasons stated in the first pages of this book, to one period and nation. But it may be remarked, that the national condition most favourable to general efficiency is one of self-confidence and eager belief in the existence of great works capable of accomplishment. The opposite attitude is indifferentism, founded on sheer uncertainty of what is best to do, or on despair of being strong enough to achieve useful results; a feeling such as that which has generally existed in recent years among wealthy men in respect to pauperism and charitable gifts. A common effect of indifferentism is to dissipate the energy of the nation upon trifles; and this tendency seems to be a crying evil of the present day in our own country. In illustration of this view, I will quote the following extract from a letter of one of my correspondents, who, I should add, is singularly well qualified to form a just opinion on the matter to which he so forcibly calls attention :—" The principal hindrance to inquiry and all other intellectual progress in the people of whom I see much, is the elaborate machinery for wasting time which has been invented and recommended under the name of ' social duties.' Considering the mental and material capital of which the richer classes have the disposal, I believe that much more than half the progressive force of the nation runs to waste from this cause."

A great deal of energy is wasted in attempting to seize more than can be grasped. There is a feverish tendency, fostered by the daily press, to interest oneself in all that goes on, which leads to perpetual distraction, and curtails the time available for serious and sustained effort. It may be worth while to mention a curious little morbid experience of my own, as suggestive of much more mischief; it is this:—A few years ago, I had foolishly overworked myself, as many others have done, misled by a perverted instinct which goaded to increased exertion, instead of dictating rest. The consequence was, that I fairly broke down, and could not, for some days, even look at a book or any sort of writing. I went abroad; and though I grew much better and could amuse myself with books, the first town where I experienced real repose was Rome. There was no doubt of the influence of the place—it was strongly marked; and for a long time I sought in vain for the reason of it. At last, what I accept as a full and adequate explanation, occurred to me; simply that there were no advertisements on the walls. There was a picturesqueness and grandeur in its streets which sufficed to fill the mind, and there were no petty distractions to fret a weakened eye and brain. When we are in health we take little count of the racket of English life, which may keep apathetic minds from stagnation, but which causes needless wear and tear to active ones, suggesting nothing useful, and teasing, distracting and wearying. I have heard German professors speak with wonder at our waste of energy in mere fidget, and in so-called amusements, which are mostly very dull, and ascribe the successful laboriousness of their own countrymen to the greater simplicity of the lives they lead; and they are a happier people than we are.

Partial Failures.—We have seen that energy, health, steady pursuit of purpose, business habits, independence of views, and a strong innate taste for science, are generally combined in the character of a successful scientific man. Probably one-half of the men on my list possess every one of these qualities in a considerable, and some in a high degree. If one or more of these qualities be deficient, success becomes impossible, unless its absence be appropriately supplemented by other qualities or conditions. Cases may be specified, in which too few of the above-mentioned qualities were present, and which consequently ended in an abortive career. One, is the possession of energy, health, and independence of character in excess, and little else to control them. These are dangerous gifts. Those who have them are apt to renounce guidances by which the great body of mankind move safely, and to follow out a career in which they are almost certain to blunder and fail egregiously. Probably every large emigrant ship takes out many such men, full of unjustifiable self-confidence, who, to use a current phrase, "knock about in the world," waste their health, youth, and opportunities, and end broken down. Another case, is that in which a strong innate taste for science is accompanied by independence of character and steadiness of pursuit, but with no other quality helpful to success, and which therefore leads to no useful result. There is hardly a village where some ingenious man may not be found who has ideas and much shrewdness, but is crotchety and impracticable. He wants energy and business habits, so he never rises. Many of these men brood over subjects like perpetual motion : their peculiarities are well illustrated in De Morgan's Book of Paradoxes. Again we frequently meet persons of a stamp that justifies the old-fashioned caricature of scientific men, who are absorbed in some petty investigation, utterly deficient in business habits, and noted for absence of mind. Even idiots have often strongly quasi-scientific tastes, as love for simple mechanism, or objects of natural history; and they have, as already remarked, a pleasure in collecting. Madmen have often persistency, as is shown by their brooding on a single topic. We all of us must have met with curious cases of failures, where a mind and disposition that promise much for success, never achieve it. It may be that some mental screw is loose, or there is some irreparable weakness of judgment, or some untimely irresolution or rashness; any fault of this kind is sufficient to mar a man's chances when competition is keen. To obtain the highest order of success, two things are wanted; first, the qualities of the man must either be good all round, or else he must be so circumstanced as to be able, when the need arises, to supplement his deficiencies by extraneous help; secondly, he must have some very useful qualities highly developed. It is said that "genius" is required for high success, and there is much talk about what genius is, and on the failures of men of genius, while some persons go so far as to doubt the existence of genius as a separate quality. It appears to me, that what is generally meant by genius, when the word is used in a special sense, is the automatic activity of the mind, as distinguished from the effort of the will. In a man of genius, the ideas come as by inspiration; in other words, his character is enthusiastic, his mental associations are rapid, numerous and firm, his imagination is vivid, and he is driven rather than drives himself. All men have some genius; they are all apt, under excitement, to show flashes of unusual enthusiasm, and to experience swift and strange associations of ideas; in dreams, all men commonly exhibit more vivid powers of imagination than are possessed by the greatest artists when awake. Sober, plodding will is quite another quality, and its over-exercise exhausts the more sprightly functions of the mind, as is expressed in the proverb, "too much work makes a dull boy." But no man is likely to achieve very high success in whom the automatic power of the mind, or genius in its special sense, and a sober will, are not well developed and fairly balanced.

CHAPTER IV.

EDUCATION.

Preliminary—Education praised throughout or nearly so—Merits in Education—Merits and demerits balanced—Demerits—Summary—Conclusion.

I NOW pass on to the education which the scientific men had in their youth, in the hope that my results may give assistance to those who are endeavouring to frame systems of education suitable to the wants of the day. What I have to say is very partial; it refers solely to the opinions the scientific men entertain of the merits and faults of their own several educations in bygone days. Their views are remarkably unanimous, considering the very different branches of inquiry they are interested in, and the great dissimilarities in their education.

One-third of those who sent replies have been educated at Oxford or Cambridge, one-third at Scotch, Irish, or London universities, and the remaining third at no university at all. I am totally unable to decide which of the three groups occupies the highest scientific position: they seem to me very much alike in this respect.

The questions to which the following replies were given, were as follows :—" Was your education especially conducive to, or restrictive of, habits of observation ? " " Was your education eminently conducive to health or the reverse ? " " What do you consider to have been peculiar merits in your education ? " " What were the chief omissions in it, and what faults of commission can you indicate ? " I also asked for information concerning the places of education, both schools and colleges, and as regards home and self-instruction. The answers were, in some cases, very interesting from their minute elaboration, but I am, of course, restricted on this occasion to a simple treatment of them. I cannot now paint with delicate tints, but must content myself with broad lights and shades. The following answers are extracts, and, in some few cases, abstracts; they convey the general tone of the several replies as nearly as possible.

The groups under which I have sorted them are these :—

Merits :—
,, Education praised throughout,
   or nearly so . 10 replies
,, Variety of subjects . 10 ,,
,, A little science at school 3 ,,
,, Simple things well taught 3 ,,
,, Liberty and leisure . 3 ,,
,, Home teaching and en-
   couragement . . 8 ,,
Merits and demerits balanced . 4 ,,
Demerits :—
,, Narrow education . . 32 ,,
,, Want of system and bad
   teaching . . . . 10 ,,
,, Unclassed . . . 4 ,,
   Total 87

There are a few cases in which an answer, already given in combination, has been extracted and repeated.

MERITS : EDUCATION PRAISED THROUGHOUT, OR NEARLY SO—TEN CASES.

(1) "Was admirably taught, æt. 13–16½, to reason, use my own mind, and depend on myself. Was taught to acquire large masses of information by reading. There was a little tendency to a vagrant style of reading, but this was probably neutralised by other influences."

(2) "Well taught in classics and mathematics. If possible my education should have afforded facilities for the study of the science of observation, but I doubt the practicability of this at school. While a schoolboy I taught myself botany, chemistry, &c., under great disadvantages."

(3) "Careful and good early education at home by my mother and father; then rather strict training by my father and by my first

schoolmaster. Being carefully looked after by my father and expected to do my best."

(4) "My education was well balanced; it was general and of a very complete kind, including chemistry, botany, logic and political economy; but 3 years (æt. 12–15) spent in learning the Latin and Greek grammars were a blank waste of time."

(5) "Education included French, German, logic, natural philosophy, chemistry, besides mathematics. I lived in a house where I saw many people whose interests were of various kinds, and I went to a day-school where I mixed with the boys only when they were fresh and active. Thus I had two outer worlds to balance against each other. On the whole, I had, I think, the greatest degree of freedom possible to a boy."

(6) "Was at school till æt. 16, and with a tutor in Germany for 6 months; after then, technical training and teaching. The education was conducive both to observation and health.

Variety of subjects and attention to details. A combination of home and school education, my father having been head master of the school."

(7) " My father being a schoolmaster, I was at some sort of school work nearly all my life, but from the age of 12 I was occupied more in teaching than in learning. My education included the various subjects usually taught in English schools, with something of astronomy, pneumatics, electricity, and mechanics. I learnt much in conversation with my father, which chiefly took an instructive form. Was led to think and speak freely, also to engage frequently in domestic discussions on questions of general policy. I had also early access to tools and materials."

(8) " I was fortunate in obtaining at school (æt. 8-16) an insight into the phenomena of nature, a subject entirely ignored at that time in almost all schools. My peculiar bent for experiment was encouraged at home by my mother, and there were peculiar merits in my

training under Professors . . . . at . . . . , and especially in Germany, under . . . ."

(9) "The steadiness with which I was taught by one eccentric schoolmaster reading and accurate spelling, clear, neat, and intelligible writing, and quick and accurate computation by all the primary rules of arithmetic. Faults in these several branches were never overlooked, and all competition was for excellence in each; Latin and French were evidently thrown in to please parents. Going to sea, at the age of 13, I really think I started with the best education I could have had. Compared with my youthful messmates, some of whom had passed through public schools, I was far their superior in writing (I soon acquired chart-drawing and sketching from nature), and in calculation of the day's work, and in astronomical observations."

MERITS IN EDUCATION: VARIETY OF SUBJECTS
—NINE REPLIES.

(1) "Not tied down to old courses of classics and mathematics."

(2) "My master (æt. 15–17) was a man of scientific and generally liberal turn of mind."

(3) "Sufficient groundwork in many subjects to avoid error."

(4) "Early introduced to many subjects of interest."

(5) "A well-balanced education [including chemistry, botany, logic, and political economy]."

(6) "A variety of subjects and attention to details. Coming in contact with persons of every rank [in Scotland], and sitting on the same form with the sons of tradesmen and ploughmen, as well as of gentlemen."

(7 & 8) Two cases; both [being Englishmen] praise Scotch system of education.

(9) "Living in a house where there were many interests, and going thence to a day-school, where there were other and different ones."

MERITS IN EDUCATION: A LITTLE SCIENCE AT SCHOOL—THREE REPLIES.

(1) "Only one good thing; that was object lessons, though given badly and only for a short time."

(2) "All the merits [of my schooling] I attribute to a little elementary physics and chemistry, taught me between the ages of 7 and 13."

(3) "Science taught me at school between the ages of 11 and 16."

MERITS IN EDUCATION: SIMPLE THINGS WELL TAUGHT—THREE REPLIES.

(1) "Clear, neat, and intelligible writing, accurate spelling, and simple computation."

(2) "Was very well grounded in arithmetic at school."

(3) "Forced accuracy of delineation at home, æt. 14–16."

MERITS IN EDUCATION: LIBERTY AND LEISURE—THREE REPLIES.

(1) "Unusual degree of freedom."

(2) "Freedom to follow my own inclinations and choose my own subjects of study, or the reverse."

(3) "The great proportion of time left free to do as I liked, unwatched and uncontrolled."

MERITS IN EDUCATION: HOME TEACHING AND HOME ENCOURAGEMENT—EIGHT REPLIES.

(1) "Encouragement by my mother."

(2) "Encouragement by my father."

(3) "Carefully looked after by my father and expected to do my best."

(4) (See (7), in "Education praised throughout or nearly so.")

(5) "During 1 year (æt. 17) I resided and studied with my uncle [by marriage] and learnt there more of the dead languages than in all my school time."

(6) "My private education at home was much the more valuable."

(7) "Home and self-education developed my observing faculties."

(8) "Pretty much self-taught, but encouraged to use my eyes, wits, and independent thought."

MERITS AND DEMERITS IN EDUCATION BALANCED
—FOUR REPLIES.

(1) " Left to myself, and I pursued a discursive line. As compared with ordinary schools, I think self-teaching has many advantages for boys of active minds; but intelligent teaching and insisting on accuracy and completeness would have produced a much more efficient man."

(2) " The merits of my education consisted in the great number of studies connected with

nature; but there was a want of system and of consecutive study."

(3) "The demerit of my education was the want of being thoroughly grounded; this gave me great trouble, but made me think for myself; often an advantage to me."

(4) "No sound instruction; the education was too general and desultory, but it gave wide interest."

DEMERITS: NARROW EDUCATION—THIRTY-TWO CASES.

(1) "No mathematics nor modern languages, nor any habits of observation or reasoning."

(2) "Enormous time devoted to Latin and Greek, with which languages I am not conversant."

(3) "Omission of almost everything useful and good, except being taught to read. Latin! Latin! Latin!"

(4) "Latin through Latin—nonsense verses."

(5) "Limitation of subjects practically to classics."

(6) "Absence of any scientific training; too much confined to classics."

(7) "Omission of mathematics, German, and drawing."

(8) "Latin and Greek were more insisted on than modern languages."

(9) "In an otherwise well-balanced education, 3 years, æt. 12–15, at a private school were spent on Latin and Greek grammar—a blank waste of time."

(10) "School work directed to the cultivation of literary tastes only, and therefore not adapted to a variety of intellects."

(11) "Elements of natural science omitted; nothing taught of the nature of the world around us."

(12) "Not taught mathematics, nor any natural science, to which I could have taken *con amore*."

(13) "Absence of instruction in the modern languages."

(14) "Want of the modern languages and of chemistry."

(15) "Want of logical and mathematical training."

(16) "Want of training in the habits of observation."

(17) "Neglect of mathematics; too much reliance on mere work of memory. Mental training overlooked in the mere acquisition of routine."

(18) "I could now wish that I had gone through at the university a good course of chemistry and physics, as a preparation for the other branches; but the main obstacle was lack of time."

(19) "Want of education of faculties of observation; want of mathematics, and of modern languages."

(20) "Not allowing my mind to follow its natural bias."

(21) "Neglect of many subjects for the attainment of one or two; not pushing mathematics to a useful end."

(22) "Not enough liberty; put back by too much grounding at Cambridge."

(23) "At school the classical education, viz., construing, parsing and learning grammatical rules, was not to my taste. At Oxford I wasted much time, having little sympathy with the university pursuits and habits."

(24) "Having so exclusively devoted myself to mathematics at Cambridge."

(25) "The classical teaching was said to be good, but I did not assimilate it. Perhaps my mental peculiarities and my special inaptitude to commit words to memory would have rendered most education, such as it was when I was a boy, ineffectual for much good. The main defect for me certainly was that *precise* verbal

memory was the test of all knowledge. No doubt, in some things, such as languages, precise knowledge of words is essential, and therefore I refer to my own special defect in saying this."

(26) " My school work was too predominantly classical, and nearly everything was taught on *authority*.'

(27) "Persistence in giving me no holiday, and overstraining my memory when I was very young."

(28) "My principal regret is that I was unable to pursue the study of mathematics."

(29) " Mathematics were not pushed far enough ; natural science was left to the boys themselves."

(30) "My boyhood was utterly wasted, and the efforts of my manhood have not sufficed, and never will suffice, to repair the loss."

(31) "Omission of all subjects excepting the classics, but particularly [faulty] in the want of intellectual training."

(32) [A military man.] "The *authority* of a military education is prejudicial to the development of thought and education in matters of opinion."

DEMERITS IN EDUCATION : WANT OF SYSTEM AND BAD TEACHING—TEN CASES.

(1) " Want of system."

(2) " Want of system."

(3) " Want of system."

(4) " Want of system ; absence of necessary control."

(5) " Bad early masters ; neglect at public school."

(6) " Essentially defective ; no competition nor supervision."

(7) " The very mistaken way in which languages, as it now seems to me, especially Latin and Greek, were taught."

(8) "Too much for memory; nothing for thought."

(9) "Want of thoroughness in early teaching."

(10) "Careless and superficial reading."

DEMERITS IN EDUCATION : UNCLASSED—FOUR CASES.

(1) "Brought up in an idle class, and never realised the necessity of labour in acquirement."

(2) "Too much cramming for examinations. Too much isolated, being the youngest son and educated at home."

(3) "Too great changes in system, having been educated at 5 universities (3 of which were Scotch, 1 London, and 1 in Germany)."

(4) "Being brought up at home; was perhaps too much shut out from the company of other boys."

SUMMARY.

The scientific men on my list have very generally ascribed high merits to a varied education. They say, as we have just seen:—"Not tied down to old courses of classics and mathematics."—"Sufficient groundwork in many subjects to avoid error."—"A well-balanced education, including chemistry, botany, logic, and political economy."—"Coming in contact with persons of every rank, and sitting in the same form [in a Scotch school] with the sons of tradesmen and ploughmen, as well as gentlemen." In contrast to this, others who speak of the faults of their education, say:—"No mathematics, nor modern languages, nor any habits of observation or reasoning."—"Enormous time devoted to Latin and Greek, with which languages I am not conversant."—"In an otherwise well-balanced education, three years were spent on Latin and Greek grammar—a blank waste of time."—"Neglect of many subjects for the attainment of one or two; not pushing

mathematics to a useful end." Evidence such as this, fully establishes the advantage of a variety of study. One group of men speak gratefully because they had it, and another speak regretfully because they had it not. I find none who had a reasonable variety who disapproved of it, none who had a purely old-fashioned education who were satisfied with it. The scientific men who came from the large public schools usually did nothing when there; they could not assimilate the subjects taught, and have abused the old system heartily. There are several serious complaints about superficial and bad teaching which I need not quote afresh. Overteaching is thoroughly objected to; thus, in speaking of merits of education, I find:—
" Freedom to follow my own inclinations, and to choose my own subjects of study, or the reverse."
—"The great proportion of time left free to do as I liked, unwatched and uncontrolled."—
" Unusual degree of freedom." There is much scattered evidence throughout the replies to my questions generally, in addition to what I have extracted, which implies that this feel-

ing is a very common one. There are many touching evidences of the strong effect of home encouragement and teaching; of this I have already spoken, and need not dwell upon afresh.

In corroboration of the conclusions stated in p. 216, on the favourable influence of the Scotch system in developing a taste for science, I remark that in these replies, a large proportion of the scientific men who have mentioned any merits in their education, were educated in Scotland.

As regards the subjects specially asked for, even by biologists, mathematics take a prominent place. Two of my correspondents speak strongly of the advantages derived from logic, and the weighty judgment of the late John S. Mill powerfully corroborates their opinions. Accuracy of delineation is also spoken of, and, owing to the extraordinary prevalence of mechanical aptitudes, I believe that the teaching of mechanical drawing and manipulation would be greatly prized.

The interpretation that I put on the answers

as a whole is as follows: To teach a few congenial and useful things very thoroughly, to encourage curiosity concerning as wide a range of subjects as possible, and not to overteach. As regards the precise subjects for rigorous instruction, the following seem to me in strict accordance with what would have best pleased those of the scientific men who have sent me returns:—1. Mathematics pushed as far as the capacity of the learner admits, and its processes utilized as far as possible for interesting ends and practical application. 2. Logic (on the grounds already stated, but on those only). 3. Observation, theory, and experiment, in at least one branch of science; some boys taking one branch and some another, to ensure variety of interests in the school. 4. Accurate drawing of objects connected with the branch of science pursued. 5. Mechanical manipulation, for the reasons already given, and also because mechanical skill is occasionally of great use to nearly all scientific men in their investigations. These five subjects should be *rigorously* taught. They are anything but an excessive programme, and there would remain plenty of time for that variety of work which is so highly prized, as—ready access to books; much reading of interesting literature, history and poetry; languages learnt, probably best during the vacations, in the easiest and swiftest manner, with the sole object of enabling the learners to read ordinary books in them. This seems sufficient, because my returns show that men of science are not made by much teaching, but rather by awakening their interests, encouraging their pursuits when at home, and leaving them to teach themselves continuously throughout life. Much teaching fills a youth with knowledge, but tends prematurely to satiate his appetite for more. I am surprised at the mediocre degrees which the leading scientific men who were at the universities have usually taken, always excepting the mathematicians. Being original, they are naturally less receptive; they prefer to fix of their own accord on certain subjects, and seem averse to learn what is put before them as a task. Their independence of spirit and coldness of disposition are not conducive to success in competition : they doggedly go their own way, and refuse to run races.

CONCLUSION.

Science has hitherto been at a disadvantage, compared with other competing pursuits, in enlisting the attention of the best intellects of the nation, for reasons that are partly inherent and partly artificial. To these I will briefly refer in conclusion, with especial reference to the very important question as to how far the progress of events tends to counterbalance or remove them.

If we class energy, intellect, and the like, under the general name of ability, it follows that, other circumstances being the same, those able men who have vigour to spare for extra professional pursuits, will be mainly governed in the choice of them by the instinctive tastes of their manhood. The majority will address themselves to topics nearly connected with human interests ; a few only will turn to science. This tendency to abandon the colder attractions of science for those of political and social life, must always be powerfully reinforced by the very general inclination of women to exert their influence in the latter direction. Again, those who select some branch of science as a profession, must do so in spite of the fact that it is more unremunerative than any other pursuit. A great and salutary change has undoubtedly come over the feeling of the nation since the time when the present leading men of science were boys, for education was at that time conducted in the interests of the clergy, and was strongly opposed to science. It crushed the inquiring spirit, the love of observation, the pursuit of inductive studies, the habit of independent thought, and it protected classics and mathematics by giving them the monopoly of all prizes for intellectual work, such as scholarships, fellowships, church livings, canonries, bishoprics, and the rest. This gigantic monopoly is yielding, but obstinately and slowly, and it is unlikely that the friends of science will be able, for many years to come, to relax their efforts in educational reform. As regards the future provision for successful followers of science, it is to be hoped that, in addition to the many new openings in industrial pursuits, the gradual but sure development of sanitary administration and statistical inquiry may in time afford the needed profession. These and adequately paid professorships may, as I sincerely hope they will, even in our days, give rise to the establishment of a sort of scientific priesthood throughout the kingdom, whose high duties would have reference to the health and well-being of the nation in its broadest sense, and whose emoluments and social position would be made commensurate with the importance and variety of their functions.

APPENDIX.

My schedule of printed questions, together with the ample spaces left for replies, filled, I am half ashamed to acknowledge, seven huge quarto pages. It would be a cumbrous addition to a publication like the present to reproduce these in the same form in which they were framed; and as the following extracts (with trifling variations rendered necessary by the change of form) cover precisely the same ground, and are sufficient for explanation, I abstain from doing so.[1]

A circular letter, in which I explained briefly the object of the inquiry, accompanied the schedule, and I

[1] I also omit the description of a notation I proposed to replace indefinite words such as "large," "considerable," because I have made no use of it in this volume. It is a modification of the class notation used by me in my "Hereditary Genius," and was alluded to and illustrated in my lecture before the Royal Institution, 1874. I have by no means abandoned its advocacy, but have learnt the necessity of explaining and exemplifying it in considerable detail before its merits and convenience are likely to become as generally recognised as I believe they deserve to be.

appended to it a reprint of a short article which I had written in the *Fortnightly Review* early in 1873, partly to show the interest with which I had pursued cognate inquiries, and partly as a guarantee of the tone and spirit in which the inserted communications would be treated. Also I presumed, and, as it has proved, not without reason, that being more or less personally acquainted with a large majority of the scientific men on my list, they would be inclined to put greater faith in my discretion than if I had been a stranger. Subject to these preparatory explanations, the following are the questions that I circulated :—

INQUIRY INTO THE ANTECEDENTS OF SCIENTIFIC MEN.

Please return this schedule at your earliest convenience, with answers to as many of the questions as you consider to be unobjectionable, and send on a separate paper any further information that you may think germane to the inquiry. Entries marked " Private " will be dealt with in *strict confidence;* they will be used only as data for general statistical conclusions.

NOTE.—Whenever you consider the grade of the quality about which a question is asked, to fall near mediocrity, *do not make any entry at all.*

Christian names of yourself, your father, and your

mother, also her maiden name? Designation and principal titles of yourself, your father, and the father of your mother? Your father and mother, are they respectively English, Welsh, Scotch, Irish, Jewish, or foreign? If foreign, of what country? Wholly or in what degree? Was either your father or your mother descended from persons persecuted for political or religious opinions, or from political or religious refugees? If so, state the precise relationship. Mention whether their political or religious opinions became traditional in the family. Occupation of yourself, your father, and the father of your mother? Specify any interests that have been very actively pursued by them, in addition to their regular occupation or profession.

All the questions in the following paragraph are asked concerning yourself, your father, and your mother respectively:—

Date of the birth of? Place of the birth of (if you do not remember that of either your father or mother, state where he or she resided in early life)? Mention if it was in a large or small town, a suburb, a village, or a house in the country. To what religious bodies have you (self, father and mother) respectively belonged? To what political parties? Health at the various periods of life? In early adult life, what was your height (to be estimated, where not accurately remembered)? Was there anything distinctive in the figure, &c. (spare, symmetrical, muscular, &c.)? Colour of hair? Complexion (if remarkably fair, dark, ruddy, pale, sallow, &c.)? Temperament, if distinctly nervous, sanguine, bilious, or lymphatic? Measurement round inside of rim of your hat? Energy of body, if remarkable; as shown by power of activity, power of enduring fatigue, restlessness, requiring but little sleep (state how much), early rising, adventures, travel, mountaineering, &c. (give a few facts)? Energy of mind, if remarkable; as shown by power of accomplishing a large *amount* of brain work, by the vigorous pursuit of interests, whatever they may be, &c. (give a few facts)? Retentiveness of memory (give facts)? Studiousness of disposition and mental receptivity, as shown by large acquirements? Independence of judgment in social political, or religious matters (give illustrations)? Originality or eccentricity of character (give illustrations)? Special talents, as for mechanism, practical business habits, music, mathematics, &c.? Strongly marked mental peculiarities, bearing on scientific success, and not specified above: the following list may serve to suggest—impulsiveness, steadiness, strong feelings and partisanship, social affections, religious bias of thought, love of the new and marvellous, curiosity about facts, love of pursuit, constructiveness of imagination, foresight, public spirit, disinterestedness.

Are any peculiarities either very uniformly developed, or also very irregularly developed among yourself, your brothers and sisters, or in the family of your father, or in that of your mother?

State the number of males and that of the females in each of the following degrees of relationship who have attained 30 years of age, or thereabouts:—Grandparents, both sides; parents, uncles and aunts, both sides; brothers and sisters; first-cousins of all four descriptions; nephews and nieces. In each of these several degrees of relationship, state the names of those who have occupied prominent positions or written well-known works, or who from any other cause may be considered as public characters. State their principal achievements, mention the best biographies, and the most useful among the scattered biographical notices that may exist of them; terms of award of medals, &c. Also, in each of the above degrees of relationship, give the number (with initials or names) of those whose ability *in any respect* was considerable, but who did not become public characters (fuller information to be sent on a separate paper). Similar information is acceptable concerning other more remote degrees of relationship. Brief notes concerning hereditary peculiarities of any kind in your family, bodily or mental, would be acceptable. How many brothers and sisters had you older than yourself, and how many younger?

How long were you at small schools, large schools, universities, and at what ages? Name or place of school or university; and chief subjects taught there. Mention any honours of importance gained by you at schools or universities. To what extent were you educated elsewhere, taught at home, or self-taught? Was your education especially conducive to, or restrictive of habits of observation? Was it eminently conducive to health or the reverse? What do you consider to have been peculiar merits in it? What were the chief omissions in it, and what faults of commission can you indicate? Has the religion taught in your youth had any deterrent effect on the freedom of your researches? Can you trace the origin of your interest in science in general and in your particular branch of it? How far do your scientific tastes appear to have been innate? Were they largely determined by events occurring after you reached manhood, and by what events?

Have you been married? Year in which you were married? Maiden name of your wife? Number of living sons and daughters (of all ages)? State any facts of peculiar interest in your wife's family.

INDEX.

ABILITY of different races, 19; ranks, 23; distribution of, in families, 72.
Adams, 8.
Adhesiveness, 194.
Ages of scientific men, 10; of their parents, 34.
Alderson, 41, 68.
Amusements, 230.
ANTECEDENTS, 1.
Aristotle, 35.
Axioms, 218.

BARCLAY, 65.
Bateman, 55.
Bell scholarship, 23.
Bentham, 8, 41, 43, 65.
Bidder, 52.
Birthplaces of scientific men, 19.
Brodrick, 65, 68, 69.
Bunsen, 8.
Business habits, 104.

CAMBRIDGE honour lists, 66, 69, 257.
Carpenter, 43.
Catholics, 127.
Charity, 228.
Clark, Miss, 54.
Clergymen, 23, 208, 259.
Clubs, 5.
Colburn, Zerah, 52.
Collections, tastes for, 194.

Colour of hair of parents, 28.
Compton, 65.
Councils of scientific societies, 24.
Creeds, diversity of, 123, 126; effect of, on research, 135.

DAGUERRE, 7.
Dalton, 124.
Darwin, 41, 45, 63, 65.
Data, 10.
Dawson Turner, 41, 48.
Definition of "Man of Science," 2.
De la Rue, 53.
De Morgan, 232.
Descent (see race and birthplace, 16).
Discovery, 7.
Divines (see clergymen, creeds, and religious bias).
Dreams, 234.
Duncan, 35, 36.

EDUCATION, 235; merits in, generally praised, 238; variety of subjects, 242; a little science at school, 243; simple things well taught, 243; liberty and leisure, 244; home teaching and encouragement, 244 (see also 205, 216, 225); merits and demerits balanced, 245;

demerits, viz. : narrow education, 246 ; want of system and bad teaching, 251 ; bad, unclassed, 252; summary, 253 ; interpretation of educational needs, 255 ; educational monopolies, 223, 259.
Encouragement at home, 205 (see also 197, 206, 259) ; of friends, 211 ; of tutors, 215.
Energy, 38, 75 ; above average, 78 ; below average, 97; wasted on trifles, 229.

FAILURES, 230.
Family characteristics, 69.
Faraday, 124.
Features inherited, study of, 40.
Fellowship, of Royal Society, 3 ; medical, 226; travelling, 219.
Female influence, 206, 211, 259 ; hereditarily, 72.
Fertility, 36 (see 102).
Figure of parents, 28.
Figures and tabulation, instinct for, 194.
Friends, influence of, 211.

GALTON, D., 47, 145 ; F., 47, 197.
Genius, 233.
Gilbert, 61, 62.
Grove, 3.

HAIR, colour of, 29.
Harcourt, 50, 65.
Head, size of, 98.
Health, 37, 99 ; of parents, 101.
Heath, 24.
Height of parents, 30 ; of scientific men, 102.
Helena, 15.
Heraldry, 197.
Herbert, 62 ; Spencer (see Preface).
Heredity, 39.
Hermia, 15.
Hill, 51.
Hinton, 61.
Holland, Sir H., 63, 100.
Home encouragement (see Encouragement).

Hooker, 49.
Humphrey, 68.

IDIOTS, 108, 195, 232 ; among elder sons, 35.
Impulsiveness, 104.
Independence of character, 121, 231 ; of parents, 122.
Innate tastes, 186 ; special, 193 ; not strongly hereditary, 196.

JEVONS, 58.

KANT, 8.
Kirchhoff, 8.
LAPLACE, 8.
Latrobe, 54, 65.
Lee, General, 55.
Leverrier, 8.
List of scientific men, 4, 6.
Logic, 255.
Love, 194.

MAIN, 6.
Map of birthplaces, 20.
Marriage, best age for, 36.
Maskelyne, 65.
Mechanical aptitudes, 124 ; drawing, 255 ; manipulation, 256.
Mechanicians, birthplace of, 19.
Medical fellowships, 226.
Memory, 107 ; good verbal, 109 ; facts and figures, 111 ; form, 113 ; good, but no particulars, 117 ; bad, 120.
Miller, Hugh, 135.
Mill, J. S., 139, 148, 255.
Milnes, 65.
Ministers (see clergymen).
Mitchell, Dr. A., 35.
Moberly, 68.
Moravians, 65, 124, 127.

NATIONAL activity, 227.
Natural ability, 227, (see also 18).
Natural groups, 2.
Nature and nurture, 12.
Nonconformists, 126.
Nurture, 12.

INDEX.

OCCUPATION of parents, 21.
Opie, 42.
Origin of taste for science, 144; extracts at length, 149; analysis of them, viz.: strongly innate, 186; not innate, 191; tastes bearing on science, 194; tastes not very hereditary, 196; fortunate accidents, 198; indirect motives, 199; professional, 202; encouragement at home, 205; by friends, 211, by tutors, 215; travel, 218; unclassed, 221.

PALGRAVE, 49, 68.
Paradoxes, book of, 232.
Parents of scientific men, their occupation, 21; physical peculiarities, 27; health of, 201; independence of character, 122; relative influence of paternal and maternal lines, 72, 197, 206.
Parker, 48.
Parkes, 63.
Pedigrees, 40.
Perseverance, 103.
Phillips, 100.
Physical peculiarities of parents, 27.
Photography, early, 7, 63.
Photographic studies of features. 40.
Playfair, 55, 65.
Plum-pudding, 212.
Politics, 207.
Population, rates of scientific men to, 10.
Portraits, 40.
Powell, B., 24.
Practical business habits, 104.
Priestly, 8, 45.
Primogeniture, &c., 33.
Prisoners, 76.
Professions, influence of, 193, 202.
Purity of type, 18, 32, 40.

QUAKERS, 65, 124, 127.
Qualities, 74.
Questions, see Appendix.

RACE, 16; ability of different, 18.
Railway statistics, 145.
Rank of scientific men, 21; as regards ability, 22.
Relatives, number of, 64.
Religious bias, 126; sects, 123, 126; creed, effect of, on research, 135.
Replies, 10; are 100 in number 11.
Residuum, the, 23; unclassed of motives, 221.
Rome, 229.
Roscoe, 41, 57.

SANDEMANIANS, 124.
Sanitary administration, 224, 260.
Scientific men, list of, 4, 6; ratio to population, 9.
School productiveness of eminent men, 67.
Scotch system of education, 215, 225, 255.
Scott, 68.
Sexual selection, 32.
Shakespeare, 14.
Smith, W., 65; Arch., 157.
Social duties, 228.
Societies, scientific and clergymen, 25.
Somerville, Mrs., 108.
Sons, elder and younger, 34.
Speciality of taste, 193.
Statistics, 147; of heredity, 64.
Sterility, 37.
Stokes, 8.
Strachey, 58.

TABULATION, instinct for, 197.
Taste for science, innate, 186; not innate, 191, (see Origin of taste).
Taylors of Ongar, 41, 60, 65.
Temperaments of parents, 27.
Travel, 218.
Truthfulness, 141, 148.
Tutors, influence of, 211.
Turner (see Dawson Turner, 48).
Twins, 13.

UNCLASSED residuum of influences, 221.
Unitarians, 124, 126.
University education, 236, 257.
Urban distribution, 19; population, 38.

VANITY, 128.

WALLACE, 7.
Watt, 45.
Wedgewood, 7, 46, 62, 63, 65.
Wesleyans, 127.
Wife, influence of, 211, (see also 207, 257).
Wilberforce, Bishop, 24.
Will, the, 223.
Woodhouse, 42, 69.

THE END.

For Product Safety Concerns and Information please contact our EU representative GPSR@taylorandfrancis.com
Taylor & Francis Verlag GmbH, Kaufingerstraße 24, 80331 München, Germany

www.ingramcontent.com/pod-product-compliance
Lightning Source LLC
Chambersburg PA
CBHW071807300426
44116CB00009B/1228